THE INVESTOR VISIT

AND OTHER STORIES

PETER TERTZAKIAN

THE INVESTOR VISIT AND OTHER STORIES

Disruption, Denial and Transition in the Energy Business

ENERGYPHILE

ISBN 978-1-9991113-0-4 (paperback)
ISBN 978-1-9991113-2-8 (ebook)

These stories incorporate fictional elements inspired by historical artifacts and are based on the author's own research. Certain names, characterizations, descriptions and scenarios are either products of the author's imagination or used in a fictitious manner.

Where deemed necessary, every reasonable effort has been made to contact the copyright holders for work reproduced in this book.

Published by Energyphile Media Inc.
energyphile.org

Produced by Page Two
pagetwo.com

Edited by Lori Burwash
Cover and interior design by Taysia Louie
Original design concept by Christina Sweetman

Contents

WELCOME TO ENERGYPHILE

OVER THE PAST few years, I've been asked many times if I'm writing another book. My reply has always been a hesitant "yes." That's because, two books later, I know better.

In 2008 I was liaising with a book publisher in New York. Having written a bestselling book already, I figured I was savvy to the process. But the advice I was about to hear radically changed my thinking.

"Just make sure the first and last chapters are good," was the directive from Debra, a publishing veteran. "Nobody reads the stuff in the middle."

I don't know why her words surprised me. As a busy finance professional, I rarely had any time to read business books from cover to cover. (I still don't.) So why did I think other people would give up several hours of their time to read my book? Nonetheless, I wrote that second book just as I'd written my first — except paying more attention to the first and last chapters that time around.

But Debra's words sat with me over the years. And when I started thinking about my next project, I knew I had to try something radically different. I knew I needed to reinvent the traditional business book. *The Investor Visit and Other Stories* is a result of that thinking.

The Energyphile Vision

The Investor Visit and Other Stories is part of a much larger project called Energyphile. Underpinning Energyphile is my collection of curiosities — things like an Edison-era light bulb, a 19th-century whaling journal, a postcard of an old oil derrick, a 1914 dictionary and other energy ephemera. Also in the mix are photographs I've taken myself from around the world. These carefully curated objects and images in my personal museum form the inspiration for my stories (like the ones you're about to read). Everything is linked together on a digital platform you can explore at energyphile.org.

Through Energyphile, you can learn about our energy past and apply these lessons to our energy present. Then there's our all-important energy future. You'll find comfort in seeing there are plenty of historical analogues from which to model go-forward decisions. Is your business being attacked by a competitor with new technology? The CEO of a 19th-century candle company has sage advice for you. How should you engage with your end-use customers? A hundred-year-old utility bill has some answers. Think government subsidies are key to transitioning our economy to clean energy? An electricity entrepreneur from 1907 may make you think differently.

For me, Energyphile has been a long time coming.

Over my two decades of writing and public speaking experience, I've always strived to engage with people about vital issues that relate to energy, environment, business and our economy. And to do it in a non-confrontational manner that — amid a climate of highly polarized opinions — gets people talking civilly, rationally.

In my articles and presentations, I routinely draw from historical anecdotes and my personal museum. Audiences appreciate the stories embedded in my energy-related artifacts, but the magic happens when a story from the past is juxtaposed with a current situation. Neurons start firing when people understand how it relates to their present-day circumstance and how it can help them think about the road ahead. Suddenly that seemingly distant tale *means something* to them.

I've seen how a story about the demise of a mining town in 1922 prompts the CEO of an energy company to think differently about how they can respond to competition and new technology. I've seen how a postcard from the 1940s inspires citizens to think about their own relationship with, and responsibility for, their energy use.

In my stories, there's something for everyone with a stake in energy (which is pretty much all of us). Use Energyphile's content to engage in lively boardroom discussions about business strategy and investment. Government is a contentious influence in energy — Energyphile can help policy wonks defuse debate. Caught in a heated dinner-table argument with relatives about fossil fuels? Energyphile's stories can help you find common ground for shaping a more sustainable world.

The Customer Is Always Right

Debra's sage advice in 2008 is even truer now. Her nobody-reads-this-stuff-anyway conjecture was the genesis for thinking about how I could reinvent the business book.

Too many words is just one problem with a genre that targets busy leaders, investors and other stakeholders. Business books usually have a half-life shorter than a new fitness fad. Charts and graphs go out of date before the pages roll off the printing press. Ideas within lose their value faster than a bad stock pick.

A decade later, the trends are unrelenting. Shortening attention spans. Rapidly accelerating change. Rapidly depreciating ideas. All against the senses-numbing din of the news cycle. Even the business of publishing and the activity of reading have been disrupted.

Before I started writing anything about energy, I knew I had to tackle all those complexities — or, better yet, take advantage of them!

As I pondered writing a third book, these considerations led me to one question: what was the most impactful way to inspire a broad audience to discuss energy issues? I knew there was a better means of communicating my ideas than doing a brain dump of 70,000 words that few people read.

So I went back to the first rule of business: the customer is always right.

Of course "right" is different for everyone, but I know that my typical "customer" wants short, engaging content that can be read in a few subway stops. They want information that's authoritative, but they don't want me to wag my finger at them condescendingly. They want information that's relevant, but they don't want it delivered via eye-glazing data and charts. In a world of anxiety fueled by brutal competition and disruption, they want the context that comes from the long, deep view — and the clarity that results. They want inspiration and guidance from those who have dealt with their challenges in the past. And if that's not enough, they want to be entertained.

Debra's words had sunk in. Instead of a traditional business book, I'd write short stories, each with timeless, thought-provoking lessons. It seems obvious in hindsight, but, as I said, Energyphile was a long time coming.

Dig into the stories in whatever order you want — each is no more than a 15-minute read. No numbers, no bar graphs, no pontificating. I promise. But there's more than enough here to digest and get you thinking and talking.

That sense of exploration is at the heart of Energyphile. If I can get you doing that, I'll consider *The Investor Visit and Other Stories* a success. Then, if you want to dive in deeper, check out energyphile.org, where you can listen to the stories, browse my museum of curiosities and discover the Energyphile network.

Enjoy!

INTRODUCTION

IF YOU THINK about it too hard (like I have for the past 35 years), you can be overwhelmed by the complexities of the energy business. From the ups and downs of financial markets and the whims of government to the care of our earthly resources and environment, it's hard to keep track of it all.

On top of that, the pace of technological innovation is accelerating. If you have solar panels on your roof, LED lights in your home or an electric car in the garage, you know what I'm talking about.

Yet those are only a few obvious, almost clichéd, symbols of our new age. Much of the high-tech disruption in energy is happening in places you can't see — far over the horizon in a distant oilfield, in the algorithms of a server room or the halls of a battery research lab.

For tech nerds like me, it's all very exciting. At face value, the trends and headlines suggest a rapid shift to a perfect Jetsonesque world of limitless clean energy.

But most vectors of change (and resistance) are not driven by technology. In the world of energy, there are strong social, environmental, governmental, business and even military forces at play. Because of these temperamental influences, which often clash, it's tough to shake our dependency on deeply entrenched energy sources like oil, natural gas, nuclear power and coal. Indeed, in this short story collection, "Hubris Defined" reminds us not to be smug about simple narratives espousing "the end of oil."

These days, energy use is rife with contention. Starting up an SUV can trigger a debate about climate change, the geopolitics of oil, social issues, the economy, carbon taxes, pipeline disputes, overconsumption and ethical concerns. Though we tend to think such issues are new, weighing energy use against societal ills has long been a prickly subject. For example, the grim post–World War II postcard from father to daughter in "When Stoke Stokes" reveals our struggles in prioritizing environment over economy. In "Nobody Tips a Scandiscope," about a 19th-century child chimney sweep, I expose the disturbing reality of human self-interest as a barrier to adopting clean technology — and eradicating a shameful practice.

As you've probably guessed, I'm inspired by, and find lessons in, what's around me, discovering parallels of disruptive change while walking on a wilderness trail or city sidewalk, looking at an old postcard, photographing a rusty piece of coal mining equipment or ruminating on a claustrophobic freeway packed with vehicles. In "Once Upon a Time...," it's children's books that inspire me to explore how our perceptions of energy and environment have evolved. In "Alfred Dickie's Utility Bill," a 1915 utility bill gets me thinking about the point at which we lost touch with our energy suppliers.

And then there are my travels, or "energy tourism," as I like to call it. Camera equipment strapped to my back, I find that my best insights often come from far away, like in the Peruvian antique store I visited in 2013. That little shop was akin to an economics lab, teaching the raw realities of how we evolve our energy use — or don't.

Inquisitive as always, I'd wondered why there were so many used irons for sale in the high-altitude antique markets of the Andes. Had some weird circumstance caused South American people to stop pressing their wrinkled clothing? Of course not, but I had to find out what was happening.

In Arequipa, an Andean mining town high in the southern flank of Peru, I walked into a congested antique store, squeezing in my shoulders to avoid knocking anything over. Down an aisle at the back, I noticed a sagging rack dedicated to irons. But these relics weren't electric. There were no cords dangling from them. No, these old Peruvian irons once burned charcoal or coal.

Pointing to the top shelf, I told the proprietor, "I'd like to look at that one," knowing full well my hand gesture spoke louder than any weak attempt at Spanish I could make. With a soft "Si, señor," he handed me the artifact.

The small brass model I'd chosen had a flourish of handmade craftsmanship. Looking at it from different angles, I wondered who the previous owner was, thinking the metallic artistry must have added a touch of import to the hot, repetitive work at hand.

Coal-fired irons are not exclusive to Peru. These flat-bottomed barbecues are still common in many parts of the un-electrified world. The principle is simple. Swing open the lid, place hot embers in the chamber, close it snug and press your Sunday best. Just be careful not to get soot on your white shirts! Oh, and don't forget to open the windows: the holes on the side and top vent out the smoke.

With the smugness of any Westerner who takes modern amenities for granted, I struggled to grasp the hassle of operating a charcoal iron in the pursuit of crisp clothing. No wonder so many of these relics have ended up at collectable stores, crammed beside typewriters and kerosene lanterns. Thank God for electricity, I thought.

The International Energy Agency estimates that almost a billion people in the world still don't have access to electricity. For them, the base living paradigm is darkness, wrinkled clothes and 19th-century coal technology. Until one day, when a utility truck arrives in the village to string up power lines and install wall plugs. Entire communities transform, and consumers run to buy electric devices. The economy grows. That's called a true energy transition, and it helps explain why vintage apparatus — like all the old irons that surrounded me — are relegated to the antique store, where people like me philosophize over them.

But the stories in this book are not just philosophy and reflection: they highlight real-world happenings (that often repeat elsewhere in history) and give us a template to think about what's needed for change, and what's important to society. For instance, in "Long Way Around," I relate how ignoring energy security in favor of near-term financial benefit can result in vulnerability of the things that power our daily lives.

Few consumer devices have evolved through *four* primary fuels. The iron is one of them. In my collection of antique irons, I can show you one that generated its heat by burning wood, one fueled by charcoal, another by gas and still others by oil products like gasoline and naphtha. Just imagine funneling gasoline into a small tank on the back of a heavy metal iron, lighting the burner inside and . . . risking a house fire. Sometimes, we make peculiar trade-offs between the safety of our energy-consuming devices and the pursuit of such trifles as vanity.

Fortunately, the one iron that's missing from my cabinet is a radioactive model. Innovation is a wonderful thing, but I'm sure the idea of loading a handheld iron with uranium was dismissed fairly quickly. So we do have our limits: some transitions are just not realistic, and safety and environment can trump vanity after all. You'll read about that in one of my favorite stories, "Rocket Man."

The transitions from irons powered by wood to coal to natural gas and oil then electricity were relatively speedy and straightforward. In fact, some societies are even leapfrogging directly from coal to electricity.

But not all transitions are so simple.

Why has it taken so long to transition from oil-powered cars to electric cars? "The Great Aha" will help you understand how factors like consumer psychology and product differentiation affect the speed with which an old appliance is ditched in favor of a new one.

Back in the antique store, I turned the brass iron in my hands one more time. This unassuming object whispered many thought-provoking tales to me. I made my way to the tiny desk near the door, where the proprietor was ready to sell. He pulled his receipt pad closer, put the carbon paper into place and wrote the invoice by hand. It was a cash sale, of course. Even though Peru has modern telecommunications, this man had decided not to invest in a wireless payment machine.

His situation reinforced that forces of change are always accompanied by friction. Put differently, there are plenty of reasons why transitions don't happen. Resistance to change is a common theme in this book. It could be financially motivated, as you'll read about in "Stairway to Hell." Or it could be as simple as denial of new technology, an all-too-common human trait highlighted in "The Investor Visit."

Identifying the Forces of Change

Understanding the evolution of our energy circumstance is tricky, largely because of the multitude and diversity of forces that can affect change. I've already alluded to many of them in this introduction. Binning the influences into one of six forces of change helps me organize my thoughts about the factors that shape our energy evolution.

Each story in this collection is dominantly associated with one of the forces listed below. Of course, forces are usually interrelated — for example, environment and policy often go hand in hand — so several stories will relate to more than one. I've identified the primary force of change for each story — see how many others you can spot as you read.

INNOVATION AND TECHNOLOGY: We can all relate to how GPS technology was a navigational step change over paper maps. Similarly, electric bulbs snuffed out kerosene lanterns and candles over a century ago. Innovation and technology are important catalysts for change in energy, but are not sufficient by themselves.

ENVIRONMENT: History has repeatedly shown that when human health and welfare are compromised by environmental degradation, we react to remedy the situation, although it often has to reach a crisis before action is taken. Typically, the forces of innovation and policy follow.

ECONOMY AND BUSINESS: Nothing goes 'round without capital. Investment is an essential force for change, and lack of it hinders transition. Competition in the free market provides the incentive to attract capital and advance the force of innovation and technology.

SOCIAL AND ETHICAL: Human issues are often either overlooked or underestimated, but they're significant vectors affecting the evolution of societal energy use. They can range from consumerism, financial self-interest and pipeline resistance to vested interests in child slave labor.

POLICY AND REGULATION: Government is a strong force. The carrot of subsidies or stick of taxation can alter how we get and use energy. Many policy tools are available, including outright bans. Unfortunately, misplaced policies can lead to unintended consequences. A heavy hand of government often triggers resistance from the force of economy and business.

GEOPOLITICS: This is a gentle term for fighting over energy commodities. A fundamental staple for all societies, energy has long been used as a lever for strategic economic and political gain. Sanctions, tariffs and expropriation of assets are common. Extreme geopolitical jockeying leads to military conflict.

Gleaning Lessons and Applying Them

Business books teach lessons. I like to impart the lessons I've learned from my career in energy; in fact, advising investors, corporate leaders and policy makers is part of my day job. So, in each story, you'll find some commonsense truisms. Tenets like "don't be in denial of change" and "the low-cost producer wins" — these are essential business mantras.

However, I've never been one to stand on a pedestal and patronize my audiences with black-and-white answers in what I know is a gray-shaded world. I respect that people bring their own experiences, knowledge and opinions to the discussions I'm trying to enable — and those discussions are richer for their voices.

More than anything, I believe in self-directed discovery that encourages critical thinking. In other words, I want you to draw out the lessons from these stories, juxtapose them with your specific situation, talk about the parallels with your peers and make better calls on the future.

Whether it's around a campfire, a dinner table or a boardroom table, coming together to share stories has always been fundamental to the human experience. That's the spirit of Energyphile and the 10 short stories in *The Investor Visit and Other Stories*. I hope they prompt much thought and lively discussion!

ALFRED DICKIE'S UTILITY BILL

INNOVATION AND TECHNOLOGY

PREAMBLE

CAN A UTILITY BILL from 1915 describe the issues that assail our energy circumstance today? "Alfred Dickie's Utility Bill" is one of the first short stories I wrote. It's also one of my favorites. By the time I was done writing, I was surprised at the richness of meaning embedded in the simple scrap of paper I'd picked up for a dollar at an antique market.

My research into the Halifax Electric Tramway Company's invoice gave me insight into how people want their energy, what prompts someone to switch to a new system (from wood to gas, in Alfred's case) and how technology reduces the cost of energy over time.

Above all, Alfred's bill helped me understand an important turning point in our energy history: the point at which we lost touch with where and how we get our energy. In turn, I realized why people today are so distrustful of companies that supply our fuels.

Alfred used to know what it took to heat his home. After 1915 he didn't. Now think about where your energy comes from. Do you know what generates the electricity behind your wall plug?

Where the gasoline in your car originates? The natural gas in your furnace? What would it take for you, like Alfred, to switch to a new energy source, such as installing solar panels on your roof?

Alfred Dickie's utility bill symbolizes much more than a $1 indebtedness to a utility company — it marks a foundational story about how new sources of energy alter societal behavior and, yes, it aptly describes the issues that assail our energy circumstance today.

A rubber stamp with a 1915 date often has a fascinating story absorbed in its faded ink.

ALFRED DICKIE'S UTILITY BILL

I'D LIKE YOU to meet Alfred Dickie. Why? Because you'll be surprised by what he can teach us about our energy choices.

I came across Alfred's utility bill at an antique fair, the type of place where small treasures are begging to tell big stories. Leafing through tattered magazines, faded sheet music and other bric-a-brac, I found a bluish piece of paper from the Halifax Electric Tramway Company. My interest was piqued when I saw the date: November 30, 1915.

A rubber stamp with a 1915 date often has a fascinating story absorbed in its faded ink. I've long been captivated by the period between 1880 and 1920. It was the glory era of energy transitions: we traded loyal horses for emotionless cars, noisy steam engines for smelly diesel motors and glowing lanterns for electric light bulbs. And don't forget the Wright brothers' first flight at Kitty Hawk in 1903. By 1915, this momentum had pushed the early adoption of electrical grids and petroleum products. With each trade-in, our lifestyles and economy changed immeasurably.

DISCOUNT will be allowed only if bill is PAID on or before TENTH of Month

ALL ACCOUNTS ARE PAYABLE AT THE OFFICE OF THE COMPANY.

Office Hours, 9 a. m. to 5 p. m.
Saturday, 9 a. m. to 1 p. m.

OFFICE: 13 LOWER WATER STREET

PRESENT THIS BILL
WHEN MAKING PAYMENT

521B

ALFRED DICKIE.
95 SOUTH PARK ST.

HALIFAX, N. S.
Nov. 30th, 1915

To THE HALIFAX ELECTRIC TRAMWAY CO., Ltd., Dr.

FOR GAS CONSUMED DURING PAST MONTH:

TO NOV. 27 METER INDEX 649 00
FROM OCT. 28 METER INDEX 639 00

MINIMUM RATE $

10 HUNDRED CUBIC FEET AT $1.25 PER M. $ 1.25

20 PER CENT. DISCOUNT IF PAID ON OR BEFORE 10th DEC. .25

NET $ 100

STOVE ACC.

ARREARS

RECEIVED PAYMENT FOR THE COMPANY

DEC. 1915

TOTAL $

PAID H. E. T. CO.

NOT RESPONSIBLE FOR LEAKAGE
Bill Must be PAID within 20 DAYS or Service will be Discontinued
CUSTOMERS MUST NOTIFY THE COMPANY OF ANY CHANGE IN ADDRESS

Alfred's 1915 utility bill gives us a glimpse into the era when coal gas was a new fuel piped into homes for heating, cooking and lighting — a luxury afforded only by the wealthy.

Turning over Alfred's bill, I nodded — drawings of a stove and a water heater confirmed I had something of value. Flipping it back, I noted Alfred's particulars: he lived at 95 South Park Street in Halifax. Having visited Nova Scotia's capital many times, I imagined his house was in the late-Victorian style: a foundation of brick or concrete shouldering a couple of stories, with a covered front porch.

Handwritten meter readings recorded that Alfred had consumed "10 hundred cubic feet" of heating fuel in November 1915. For this, he was obliged to pay $1.25 "within 20 DAYS or Service will be Discontinued," according to the rather terse Halifax Electric Tramway Company.

Thankfully, he paid within three days. His speedy December 2 payment, acknowledged by a PAID rubber stamp, earned him a reward of 25¢. Nice, I thought: pay early and earn a 20% discount. These days the mentality is reversed: pay late and earn a penalty. I wonder, When in our energy history did the attitude shift from carrot to stick?

A buck went a long way a hundred years ago, so I didn't know if Alfred's bill was cheap or expensive for that brisk Atlantic month of November. Regardless, his payment of $1 seemed fitting because, a century later, my antique dealer was asking the same amount — for a piece of paper most people would have thrown out decades ago. Not me, of course.

As I flipped the bill over a few times, I pondered. Who was Alfred Dickie and what could we learn from the decisions he made a century ago? What was the Halifax Electric Tramway Company and why was it so keen to promote gas? Relative to today, was $1.25 a little or a lot to pay for gas?

I looked up and said, "I'll take it."

The soppy-faced antique dealer accepted my dollar without hesitation. That dollar bought me a treasure trove for my curiosity.

The Lumber King

I had occasion to go to Halifax in 2014. Alfred's bill in hand, I found some answers. South Park Street is still there, green and leafy as it doubtless was in 1915. However, Number 95 is no longer, most likely demolished during construction of the Victoria General Hospital.

Alfred Dickie was a prominent Canadian businessman, nicknamed the Lumber King of Nova Scotia. After attending Dalhousie College in Halifax, he returned home to Stewiacke and went into the lumber business. He wasn't the first to see value in Nova Scotia's forests, but historical accounts suggest he was one of the best at harvesting the renewable resource.

Alfred's operations grew rapidly. He bought up three large forest properties (18,000 acres here, 65,000 there) and a number of sawmills and launched several pulp and lumber companies. He even owned a few steamships. Soon, the Lumber King became the leading exporter of Nova Scotia forestry products.

But his fortunes didn't last. Like so many capitalists in a volatile commodity business, Alfred overextended himself. Bad luck — fires and harsh weather, aggravated by a lack of insurance — and a tough business cycle contributed to his downfall. The banks took over and Alfred lost all but one share of his company's equity.

Despite his misfortunes, he remained a director of his holdings... and was evidently paid well to do so. In the early 1900s, his salary was recorded as $833 per month. Now, $800 doesn't sound like much but, adjusted for 100 years of inflation, Alfred's annual income today would have been almost $250,000 — not bad for a bankrupt CEO. He was hardly suffering financially and, like any determined entrepreneur, he worked to recover his wealth and station. In 1912, Alfred went into semi-retirement and moved into that Victorian house on South Park Street.

Knowing that Alfred was wealthy through boom and bust gave me valuable context for his $1.25 gas bill. I discovered that while he was earning his $833 monthly paycheck, those Nova Scotia workers who sported overalls and lunch buckets earned 25¢ an hour — less than $45 a month. While a dollar for piped-in gas was loose change from Alfred's pocket, it would have been a considerable chunk out of the wallet of any hardworking fisherman, bricklayer or carpenter.

But the numbers weren't adding up.

I returned to the amount of gas — "10 hundred cubic feet," or 1,000, as it would be written today. A modern North American household with a furnace, water heater and stove consumes between 8 and 10 times that much gas. Alfred was either miserly with his energy use (particularly so in the cold month of November) or his house had only one or two small gas-fired appliances. The latter made more sense.

According to the bill, the Halifax Electric Tramway Company offered "Gas Appliances of all kinds." Because Alfred's usage was so low, I suspected he'd installed either a stove or a water heater. Neither appliance consumed much compared to a gas furnace heating an entire home. In fact, gas furnaces were not common in 1915 because most homes were still heated by burning wood. So Alfred's low-volume utility bill represented an early stage of transition to the gas era in Halifax.

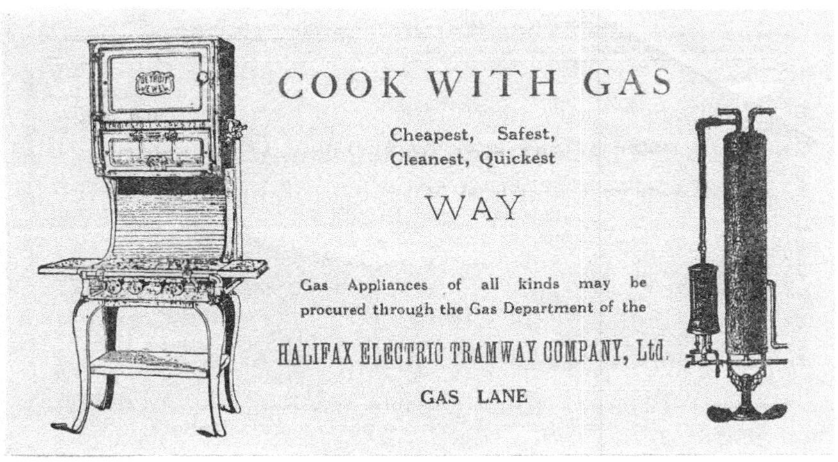

The other side of Alfred's bill advertised the superlative virtues of coal gas — and promoted some appliances while it was at it.

Alfred's low-volume utility bill represented an early stage of transition to the gas era in Halifax.

Now We're Cooking with Gas!

The history of the Halifax Electric Tramway Company is intriguing. This contemporary energy utility came into being through a series of corporate consolidations. Three entrepreneurs — Benjamin Franklin Pearson, Frederick Stark Pearson (no relation) and Henry Melville Whitney — foresaw opportunity in providing citizens with alternatives to chopping wood, riding horses and burning candles. What an incredible time to have business acumen. I imagined the trio's excitement as they realized they could make a fortune converting Halifax to the modernity of trams, electrical appliances and heat from gasified coal.

Thanks to their foresight, Pearson, Pearson and Whitney did indeed get rich, and no wonder. Think about what they were offering: instead of going outside in the cold, finding your ax, chopping a tree trunk and hauling in the wood (which would take several minutes to light and still longer to give off heat), you could simply turn the gas valve and flick a match. Today we take the luxury of instant, hassle-free energy for granted. Yet, in 1915, it was life-altering. A gas flame wasn't just faster and more convenient; it was clean and touted as safe, without the nuisance of a wood fire's smoke and leftover cinders.

With that lure, the well-to-do citizens of Halifax had good reason to hop on the city's electric tram and travel down to "Gas Lane," where they could visit the small showroom of the Halifax Electric Tramway Company. I could just imagine a young sales buck in that showroom, coaxing families like the Dickies into installing a gas line and buying a shiny new cast-iron stove. The confident salesman probably added, "May I interest you in a water heater too, sir?" Seeing the benefits, the Dickie household must have been easily convinced to switch from wood to gas, and presumably from lanterns to light bulbs too.

In fact, the marketing claims on the back of Alfred's bill were compacted into four superlatives: "Cheapest, Safest, Cleanest, Quickest." Not "Cheap, Safe, Clean, Quick," not even "Cheaper, Safer, Cleaner, Quicker." No, the Halifax Electric Tramway Company's gas appliances apparently topped all other energy alternatives.

It goes without saying that consumers still want energy to be the cheapest, safest, cleanest and quickest. We want affordable energy that doesn't pick our wallets, safe energy that doesn't harm our person, clean energy that doesn't degrade our planet and quick energy that doesn't test our patience for warmth, light or transport at the flick of a switch or turn of a key.

"Cheapest" has become even cheaper for people like you and me. Adjusting for inflation, we pay about one-third for our gas of what Alfred paid for his. The delivery of our gas is safer and more reliable too, especially after most utilities migrated from coal gas to natural gas in the 1930s.

My attention turned to society's present-day situation: if, a century ago, we had already transitioned to the "cheapest, safest, cleanest, quickest" energy source, where does that put us today?

I think we can agree that consumer attitudes have hardened considerably since Alfred's bill. In the face of unsettling issues like climate change, environmental protection and not-in-my-backyard resistance, the list of superlative adjectives we require in our energy sources is growing. Now we're looking for such claims as "lowest carbon," "fewest pipelines" and "greatest social conscience."

That thought left me unsettled. I put Alfred's bill on my desk and stared out the window. Power lines, street lamps and cars were unavoidably in my view. Surely, the list of required superlatives for our energy can't grow ad infinitum, or even beyond the length of what can be advertised on a simple energy bill. Yet why do we seem to think the list can grow?

Trading Blisters for Convenience

Picking up that blue paper again, I realized something big had happened when the people of Halifax transitioned to direct-to-the-home energy services. Utilities like the Halifax Electric Tramway Company coaxed citizens into surrendering a primal survival chore: going out and getting their own supplies.

When someone like Alfred Dickie chopped his own wood, brought it into his house and lit his kindling, he was in control of his energy, from the wood pile to the stove. He knew if he was running out of logs, if his fuel was not cut to size or if it was too wet to burn. By necessity, people of his era had a vital connection to their energy sources, especially wood. They knew how to use it wisely and were acutely aware of changes in its availability. Any incompetence or oversight could lead to cold meals at best, a family freezing to death in the middle of winter at worst.

Maybe I'd stared at the bill too long, but I had a revelatory thought. The clerk who stamped Alfred's bill with a dull rubber thud marked a symbolic moment, when consumers and suppliers lost a shared sense of responsibility for society's energy needs.

Over the last century, switches, valves and thermostats have made our lives convenient. But those knobs and dials have also pushed energy sources "upstream," out of our homes, and out of our minds, to distant horizons where pipelines and wires terminate. To be sure, separating source from usage has led to comfort and prosperity. But when we ceased getting blisters from a sweaty ax handle, we unwittingly became detached from the complexities of getting heat and light at the flick of a switch.

From the suppliers' perspective, I note the lesson of Alfred's bill in the difficulties today's energy companies face with securing approvals to build things like pipelines, power plants or even wind farms. Sure, they are all regulated into supplying the safest and cleanest according to government specs. And each competes against the other to be the cheapest and quickest. But whatever they do, companies discover that being the best is still never good enough for the most demanding fringes of the populace.

If, a century ago, we had already transitioned to the "cheapest, safest, cleanest, quickest" energy source, **where does that put us today?**

What will narrow this functional imbalance between suppliers and demanders?

Okay, I admit that the gap to achieving "best" is vague. That's intentional, because I can't predict what will satisfy everyone's needs, short of a perpetual motion machine that delivers free energy, forever, with no pollution.

I do know that a realistic "best" offering will have to come from energy suppliers and consumers rebuilding a common mindset. We don't live in a world where the driver of an SUV has to work on a drilling rig to bring a barrel of oil out of the ground. But we can live in a world where oil and pipeline companies interact better with consumers, giving the person behind the steering wheel a stronger grip on the complexities of getting gasoline to the pump. Such interaction would benefit any company, government or investment group trying to introduce or build a new energy system, renewable or otherwise.

Come Together

At the start of this story, I suggested you might be surprised by what we can learn from Alfred Dickie's utility bill. Alfred chose to transition from his wood-based system to the latest source, gas. By signing up with the Halifax Electric Tramway Company, he outsourced his barehanded duties and gained convenience.

Today, we take that convenience for granted, and have all but relinquished our ability to provide our own energy. While we don't need to start chopping our own wood again, I believe that to make truly responsible choices in the future, we must all — consumers and suppliers — reacquaint ourselves with the challenges of demanding and providing superlative-laden energy.

From my career in energy, I know it's not realistic for consumers to take a crash course in supply-side economics. I also know that most suppliers have lost touch with their end-use customers. Nevertheless, where there are business problems, there are winners waiting with solutions.

I made another business trip to Halifax in July 2015, almost a hundred years to the day after Alfred's utility bill was issued. I had some time before my flight home, so I rented a car and drove an hour north to the Pine Grove Cemetery in Stewiacke, where Alfred was laid to rest in 1929. A large personality in life, he accordingly had the tallest gravestone in the cemetery. The golden-tinged granite seemed to insist that he'd died a wealthy man.

Ironic, I thought: someone who'd made so much money selling wood — the Lumber King himself — was a lead adopter of a fuel rival. That salesman on Gas Lane must have been pretty good at gaining Alfred's trust. That, and having the lure of cheapest, safest, cleanest and quickest to back up his pitch.

It just goes to show that those who understand energy superlatives, and how to build trust around them, will be the leaders of the future.

FACING In death, as in life, Alfred looms large.

DIED
SEPT. 6
AGED 69
HIS WIFE

ROCKET MAN

SOCIAL AND
ETHICAL

PREAMBLE

I WAS A NERDY KID who grew up fascinated by space exploration, rockets and the promise of flying cars. Reading magazines like *Popular Mechanics* and following Apollo space missions were more than hobbies, they were daydreaming rituals.

As a kid, I'd never heard of Max Valier, one of the lesser-known pioneers of rocketry and space. Later in life, when I read about him in a 1929 issue of *Motor Life*, it took me back to my childhood reveries. So I had to write a story about him.

Max's antics appeared to be more showmanship than leadership. His rocket-car experiments couldn't possibly have led to inexpensive, or even safe, modes of transportation. But as you read his story, think about it in the context of technology, management and change, then consider the people whose innovations have altered our world: Thomas Edison, Steve Jobs, Bill Gates, Elon Musk, to name a few. What character attributes have driven them to deliver disruptive yet pragmatic products?

To what extent are substance, risk-taking leadership and confidence criteria for changes that alter the course of society?

When it comes to entrepreneurs who may champion the next blockbuster product, it's always tough to gauge the right balance. In my job, I routinely engage with leaders dreaming to make big changes. Many have hard-to-believe business ventures, and that's okay. To be honest, if they don't have passionate aspirations, I quickly lose interest. But many more have only bravado without the substance to back it up — that's a non-starter.

"Rocket Man" will take you to the extremes of risk, self-assurance and wild ideas, and it will get you thinking about the character attributes needed to materially shape history — or at least justify a good investment.

Max Valier was selling dreams — in a sense, pre-ordering the future.

ROCKET MAN

LIKE MANY PEOPLE who grace the pages of motoring magazines, Max Valier died a fiery death at a young age doing what he loved . . . but it's not exactly how you're imagining.

I first learned about Max in a long-forgotten magazine called *Motor Life*. He never lived to become a household name, but I think his limited fame deserves recognition beyond his native Austria. You see, Max can teach us about a prerequisite for changing the world: bravado.

Max was a space enthusiast first and foremost. Some of his pioneering rocket experiments from the 1930s inspired the next generation of brilliant scientists and engineers. I'm sure more than a few mid-century innovators tipped their slide rules in deference to Max as they figured out how to put boots on the moon and launch satellites over our heads.

But he also inspired people to think differently about mobility here on planet Earth. In the early 20th century, while everybody was talking spark plugs and gears, Max was working on rocket-powered travel.

Max was a colorful physicist, garage inventor, author and visionary — not necessarily in that order. Judging by his public antics, I imagine him to have been a suave fellow, a macho intellect who might have shown up at an art deco salon in Berlin and tossed back schnapps as he talked about his then-popular book, *A Daring Trip to Mars*, charming an admirer with a sly "It's not rocket science, you know."

My copy of *Motor Life* was published in April 1930. Back then, in the golden age of Ford and Mercedes-Benz, Max was thinking of how to power vehicles in ways other than the already-entrenched internal combustion engine. Noted for his "many rocket experiments" and for having an imagination larger than the Austrian Alps, he must have stripped the mental gears of his peers with proposals like "Why not strap a rocket engine to four wheels, gas it up with liquid oxygen and light the burner?"

Max championed several sleek rocket-car designs, many of which were showcased to curious Europeans eager for novelties. I wish I could have been at one of the carnival-like demonstrations of his controlled explosions on wheels — with him in the driver's seat.

Smiling at the photos of Max's crazy, heroic stunts, I cast myself back to an April day in 1930, when a crowd of excited men, women and children had gathered at a village racetrack in Germany.

FACING Max Valier's liquid-oxygen rocket car married his two passions — space propulsion and cars — and his vision helped usher in the age of space travel.

A HUSH SETTLES over the anxious spectators filling the wooden bleachers. Max, shoulders straight, strides toward his sleek car. As he waves to his rapt audience, he flashes a grin to a pretty woman sitting up front. Instinctively he pulls down on the lapels of his pinstriped suit and adjusts his tie, timing his vanity protocol to end just as he steps up to his rocket car. He lifts his right leg, pausing momentarily to ensure people see his two-toned shoes, then climbs into the tight cockpit. This veteran rocketeer knows that his choreography validates the unwritten rule of 1930s showmanship: swagger trumps safety.

"Stand clear!" Max shouts to his pit crew as he hunches forward and turns valves under the steering wheel.

Unbeknownst to his audience, he breaks into a mild sweat as he hears liquid oxygen swish through the pipes leading to the engine behind him. Max knows that any leak and spark combination will turn the entire car into a giant match, with him at the tip.

Sitting up, Max takes a deep breath as he releases the safety brake. The wheels move a few centimeters. The technicians scramble away. Max sucks in another deep breath and, with a twist of his wrist, flips the ignition switch on the thin dashboard. Eyes instinctively closed, he waits only a split second before the rocket burner lights up behind him, vibrating his body back to front.

Roaring heat, light and smoke spew out of the engine's raucous exhaust pipes. In the bleachers, the tension strains every human sense. Dogs start barking uncontrollably, though nobody notices above the roar of the crowd and rocket engine.

Sitting up front in the stands, pencil and notepad in hand, the *Motor Life* reporter scribbles a quick understatement: "The car is set in motion by the force of escaping gas from three steel tubes attached to the rear of the driver's seat."

There was method to Max's madness.

The car is now beginning to creep, restraining itself from the surges of the flaming force. A few seconds go by before Max turns the wheels down the straightaway, releases the brake and opens the fuel valves a full turn. Gaining momentum, the rocket car suddenly bursts down the track, to the crowd's delight. Collectively, they snap their heads and follow the trail of white smoke into the distance.

Max lives up to his hero reputation, again.

Pre-ordering the Future

It's safe to say that anyone who witnessed Max searing a racetrack knew that his bonfire on wheels wasn't a practical mode of public transport. The spectacle was undoubtedly thrilling to watch, but no sane person was going to fire up one of these things to run chores or drive down a village *Hauptstraße* to get pretzels from the corner grocer.

But there was method to Max's madness. I don't think he was expecting his unrealistic cars to replace the luxurious ride of a Mercedes-Benz roadster. He was selling dreams to his audience, showing them the possibilities — in a sense, pre-ordering the future.

Max's antics were similar to those of other charismatic techno-celebrities who changed the course of history. Big-minded titans of innovation like Thomas Edison, Henry Ford, Nikola Tesla and Steve Jobs were all masters of showcasing — if not showboating — their fantastical inventions to the public.

Among more contemporary tech visionaries, it strikes me that Elon Musk, the legendary founder of future-looking companies like SpaceX and Tesla, is an uncanny double-take of Max Valier. Both share a love of invention, space travel and new-age mobility. Both share a love of public theatrics too.

Musk's bravado in pitching things like electric vehicles, trips to Mars and hyperloops tickles the imaginations of investors, engineers and tech-hungry consumers. His self-confidence, onstage flair and closely followed social media posts allowed him to raise billions to finance history-bending ventures.

Here's the thing: much of what Musk began pitching circa 2010 — his dreams about electric cars and space adventures — was nothing new.

Bring on the Bravado

To corroborate my déjà vu, I pull out my February 1971 issue of *Popular Science*. Page 51 pitches readers on "How We'll Search for Life on Mars." Flip to page 54, and the cover story tries to convince us that "New Electrics Make Performance Breakthroughs."

Much of what Musk began pitching circa 2010 — his dreams about electric cars and space adventures — was nothing new.

(Funny how visions of space travel to Mars and futuristic cars seem to appear in pairs.)

The article claims the sleek-looking 1971 Voltair, the magazine's cover darling, delivered good range and speed. Yet, despite its supposed breakthroughs, this all-electric vehicle was never adopted into mainstream markets. Why? I'm pretty certain the high cost of the Voltair's "tri-Polar lead-cobalt batteries" had a lot to do with its dismal sales: in 10 years of production, only *60* were purchased. Maybe if Robert Aronson, CEO of the Voltair's manufacturing company, had taken a page from Max's playbook, he might have garnered more attention, and sales, for his car.

So electric vehicles have come and gone in the last century, but none gripped the imagination of the public until Elon Musk and his Tesla models came along.

There's no question in my mind that panache, audacity, bravado — call it what you want — is a requirement for bringing disruptive products to market. And the scale of the bravado required goes up exponentially with the magnitude of the disruption. If you want to change out a billion gas-powered vehicles in the world, you'd better be able to sell full-color dreams, not just dull science experiments and spreadsheets.

What Max and other trailblazers teach us is that while technology inevitably changes the course of history, its champions' passion, bravado and showmanship will bring on the change much sooner.

Sadly, Max's trail was cut short prematurely. In May 1930, one month after the publication of my *Motor Life*, he died while running tests of an alcohol-based rocket fuel. Bolstered by two successful tests, he decided to run another. With that third attempt, Max became the first casualty of the modern space age. He was only 35.

Max may not be famous today, but his work was the foundation that others built on. For example, in 1927, he cofounded the Verein für Raumschiffahrt, or Society for Space Travel. Max's dreams and experiments were inspirational to those who made spaceflight a reality in the 20th century.

I wonder where history would have taken us if Max Valier had lived longer. Maybe we'd all be driving rocket-powered cars.

THE INVESTOR VISIT

ECONOMY AND BUSINESS

PREAMBLE

YOU'RE ABOUT TO MEET two of my favorite characters: Fred, the incorrigible stock promoter, and Allan, the flamboyant oil CEO. I enjoyed penning their personalities because I've met so many Fred and Allan types in my career — the ever-optimistic, colorful personalities who push investment ideas. In their world, the pot of gold is much closer than the rainbow. And, of course, money can never be lost.

Thanks to the Freds and Allans of the world, we've all lightened our wallets on a bad investment. I'm no exception. But I've been managing money long enough to be a trained skeptic and have seen a lot of money go down the drain, proving the adage "If it's too good to be true, it probably is." That's not just in the oil and gas business, where this story is set. Fast talkers everywhere love transitional periods, when dreams of the future are easy to sell amidst change.

I've also had the pleasure of learning from characters like Graeme — sage investors who recognize the importance of gauging risk, assessing costs and being on top of fast-changing trends. Those whom we call "shrewd" always have these three antennae fully extended.

So "The Investor Visit" is about more than the characters you'll find in business. The 1915 photograph of an oil well that inspired this story unearthed solid lessons that — over a hundred years later — still stand the test of time.

Who were these people in the photo and **what brought them there that summer day?**

THE INVESTOR VISIT

"NOT LONG NOW! It's just over this hill!" Fred exclaimed. Even against the car's noisy engine and grinding wheels, the stock promoter's excitement was apparent.

"I hope your hunch is right, Fred!" Graeme yelled as he gripped the shuddering steering wheel of his late-model Chalmers-Detroit. "I'm already sold on the future of oil, but out here in this wilderness?!"

"I think you'll be impressed with the operation, Graeme," Fred replied confidently.

It was a glorious summer day in the Canadian Rocky Mountains. Graeme had taken off the car's top, unrolled the windows and removed its front windshield. He'd figured they may as well enjoy the ride before they got down to business.

Sitting beside their respective husbands, Amy and Esther clung to their hats lest they blew away. They'd expected a relaxing day trip in Waterton Lakes National Park, not this dusty drive to a drilling rig. Regardless, they were determined to make the most of it.

Suddenly, Fred lurched forward from the back seat. "There's the rig now!"

The Original Discovery Oil Company operated this site near Waterton, Alberta, from 1914 to 1920. To the left of the rig is a car full of visitors. In the bottom right, two men who likely worked there.

Graeme kept his eyes on the last turn in the road, but Amy and Esther looked to where Fred was pointing.

"Well this certainly isn't beautiful Waterton Park, is it, Graeme?" Amy muttered to her husband.

As they crested the hill, Graeme surveyed the discouraging view himself. In the valley below, a tilted smokestack belched black clouds and barrels were strewn around the site.

Fred giggled as Graeme maneuvered his car to the rig, cranked the safety brake and cut the engine with a flourish. He'd used the gains of his last two oil investments to buy it and was still smug about the purchase.

Up ahead, a man waved at them.

"That's the photographer," Fred explained, hopping out of the car. "He's photographing the property to show other investors how their money's workin' for them. Let's give him a handsome shot."

With that, Fred crossed his arms and struck a pose. The others remained where they were, faces grim, trying to ignore the dull pounding of the drilling rig's steam engine.

After tensing their muscles for what seemed an eternity, the group was relieved when the photographer sprung from behind his black velvet camera hood. Hands cupped to his mouth, he shouted, "Thank you! All done!"

Thus, a fleeting moment in the summer of 1915 was captured for posterity.

Zooming in on a Classic

I found the photograph of these four day-trippers at an antique fair in the early 2000s. The print was in rough shape. Mounted on black cardboard that had obviously had a decades-long love affair with dust, it was poorly exposed and cracked around the edges.

Nonetheless, nostalgia drew me to that image of the wooden derrick perched awkwardly in a southern Alberta valley. Recognizing a classic energy vignette, I bought it immediately.

Years later, I made a high-res scan of the print and spent hours restoring it. As I zoomed in to every nook and cranny, I reflected on the environmentally challenging scene — the barrels littering the site, the coal-black smoke drifting down the valley.

In contrast, the well-appointed visitors look out of place, dressed more for a picnic in a Henry James novel than a visit to this drilling rig in the middle of nowhere. Who were these people and what circumstances brought them there that summer day?

"GRAEME," SAID FRED, "you made your fortune in Alberta coal, but I'm tellin' you the future's oil. Every wheel in the world is gonna turn with the stuff. Just wait, you'll have a smile from here to Texas once this here well starts gushin'."

Years of investing had made Graeme shrewd. Recognizing the shift to folksy patter, he walked a few paces before responding to Fred's rehearsed hyperbole.

"I already told you, you're preaching to the choir." As if to underline his point, Graeme kicked a rock. "I know more folks are getting rid of horses and buying automobiles. Meaning more lubricants, more oil for fuel. Why the devil d'you think I bought shares in this company you're promoting?"

"You got it, Graeme!" Spying an opportunity, Fred pounced. "I sure wouldn't wanna be a chump raising horses now. Or how about making horseshoes?" He shook his head and whistled through his teeth. As they approached the great steam engine, Fred continued his sales pitch. "Besides oil and autos, the future is gas-pumping devices, stations to cater to travelers." He added cryptically, "You might be interested in other deals I've got cookin'..."

Between Fred's promotional babble and his own first impressions of the site, Graeme started to feel queasy about his investment. He was relieved to see Allan, the CEO, striding toward them. Surely this man would ease his concerns.

You couldn't miss Allan. His crisp linen suit, silk cravat, black shoes with beige spats and straw boater made him look like the gent who guessed your weight at the carnival. Graeme recognized the flamboyant attire as a common status symbol for nouveau riche oilmen. A little flashy for the investor's taste, but he didn't mind — he knew guys like Allan were tough enough to use gasoline for aftershave. Graeme looked for passion, not fashion, in a CEO.

"Quite the operation you got here," Graeme greeted Allan, noting the executive's firm handshake.

"Before you know it, this well here will be gushin'!"

Allan sounded a lot like his pitch man, Graeme noted skeptically.

The engine's banging metal, creaking wood and puffing steam distracted the threesome for a moment, but they snapped out of their reverie when the rig hand stepped from behind the shed. In grease-stained flannels and a moth-eaten hat, he was the only person who looked like he belonged on the site. Allan waved him over.

Graeme looked for passion, not fashion, in a CEO.

"Fellas, this here is Jake. He's the one who really runs the operation." Allan chuckled.

As Jake reached out to shake hands, a grinding squeal sounded from the engine. Letting loose a stream of Anglo-Saxon, he swiftly turned and yanked some levers, oblivious to the grease splattering his shirt.

For Graeme, the sight of Jake wrestling the rig works was an entertaining contrast to Allan's Rockefeller act. The term "tough guy" didn't do Jake justice. Graeme could easily imagine this foul-mouthed driller smoking a cigarette rolled out of coal.

Fred, Graeme and Allan started up the hill, away from the noise.

"We're planning to drill at least a thousand feet, Graeme," Allan declared. "Thirteen years ago, wells around here were flowing three hundred barrels a day from a zone starting at one thousand twenty. At the rate we're pounding out the rocks, we should be at target in" — he clicked his tongue as he made the calculation — "about a month."

Fred, spotting another chance to sell, jumped in. "Notice the rig is burning coal in the steam engine? It's cheap and plentiful. We pick it up from nearby mines. It might even be comin' from one of yours, Graeme." He grinned and winked.

Put off by the over-familiarity, the visiting investor glared at both men. "Can't say I'm thrilled. You're using my fuel to explore for a competitor," he said matter-of-factly. "But I'll adapt and make money on both ventures."

Fred and Allan nodded in support. They knew why Graeme didn't find this funny. Diesel engines were starting to take over from coal-fired steam engines, especially on ships and trains. Graeme's coal mines were vulnerable.

"Listen, I'm not scared of change. I'm not scared of investment risk. I love both. Put them together, you can make a lot of money. But what kind of two-bit operation are you running here?" Graeme shrugged. "Why use this old cable-tool rig that bashes the rock with a spike? If you had a more modern setup, you could drill *two* holes with my money, instead of just one."

Allan's cheeks flushed under the criticism. "This site —" he blurted.

"— landed in your lap," Graeme interjected. "You got the grandfathered drilling rights, and you're kidding yourselves if you think those'll last forever. Hell, you could lose 'em tomorrow if you keep making this mess." Graeme stabbed a finger toward the carelessly strewn barrels. "I don't want to lose my money because of more government regulations."

Graeme's voice notched up along with his irritation. "Besides, word is, there are better prospects north in Turner Valley." He paused, hoping for a response, but when none came, carried on. "I don't see any opportunity to scale here. This is a national park!"

The CEO's cheeks grew redder.

Fred cut in before Allan spoke his mind Jake-style. "You're right, Graeme. Rotary drilling *is* the next big thing. But this rig was ready to go, it's paid for, and we'll get the job done. Once we're producing oil, we'll have cash. We bring in fancy new tools — costs go down, production goes up! And we'll clean this place up, right, Allan?" He patted Allan on the back. "We'll get a fellow on the Turner Valley play. Winner takes all, Graeme. You can't say fairer than that?"

The investor could see he wasn't getting through. Fred and Allan forced thin smiles as he gestured dismissively at the rig.

"Here's the rub: I get annoyed when I see my money funding high-cost, antiquated operations that'll go the way of the candle. Whether it's drilling for oil, or mining for coal, or building horse carts or... windmills!" He hauled back and kicked at another rock. "Doesn't matter if you're in yesterday's tired business or have your finger on the doohickey everyone'll be using tomorrow. You've got to lead change, drive down costs. If you don't, you'll lose a lot of money. A lot of *my* money."

Graeme's words echoed through the valley.

Allan cleared his throat. "We're on it! You have my word, sir." Brushing dust from his spats, he offered, "Why don't I give you a proper tour?"

As they walked back down, Allan pointed out the rig mast and began outlining their grand plans for the operation.

Graeme had to admire their optimism, but his lecture had obviously fallen on deaf ears.

“**What kind of two-bit operation** are you running here?”

One Great Big Opportunity

It was fun to make-believe Graeme, Fred, Amy and Esther, not to mention dapper Allan and salty-tongued Jake. The real identities of the people in the photo are anyone's guess, but, given what I know about oil and gas finance, I'm probably not far off on their roles.

Besides, my story isn't complete fiction.

In the late 19th century, entrepreneur John Lineham saw oil seeping from the ground by Cameron Creek, which cuts through the spectacular southern Alberta valley close to where the photo was taken. Raising money and mobilizing equipment, Lineham's Rocky Mountain Development Company drilled its first well in 1902. At 1,020 feet, its rig struck oil and produced a very respectable 300 barrels a day — at first.

The promise of riches, from the original well and those that followed, evaporated quickly. Production from the company's last well, drilled in 1905, withered. Other companies had similar lack of luck in the area. Oil City, the resource town that had started to develop around the boom, was abandoned.

However, dreams of the region's big oil gushers didn't fade easily.

A decade later, the Original Discovery Oil Company chased those dreams with hyperbole. Chalking up earlier failures to Rocky Mountain Development Company's lack of precaution and funds, Original Discovery claimed in its prospectus that it was "in a position to offer to the shrewd and careful speculator **One Great Big Opportunity**." The recently opened Alberta Stock Exchange was a boon for high-risk companies and stock promoters like Fred — in 1914, Original Discovery raised a

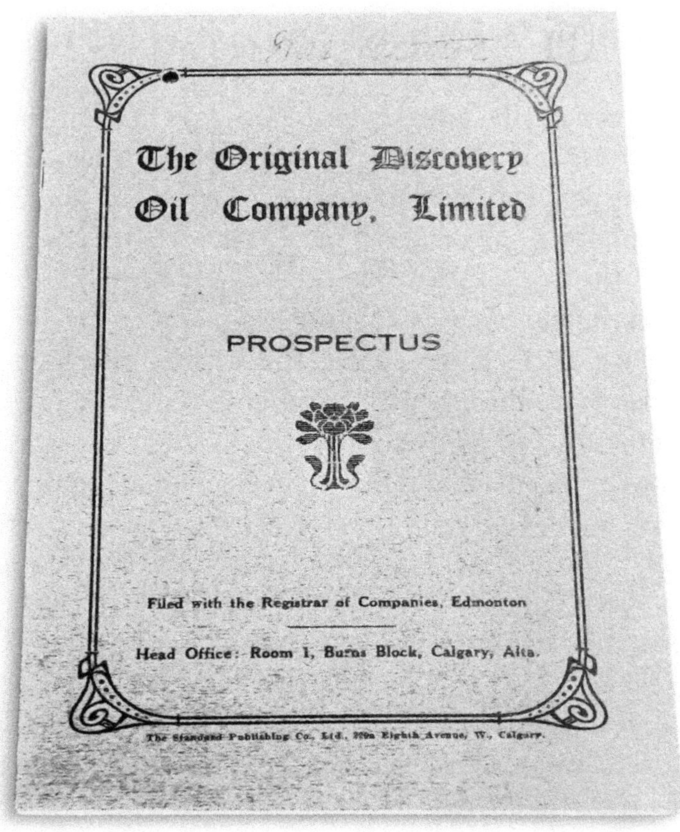

The Original Discovery Oil Company, Limited

PROSPECTUS

Filed with the Registrar of Companies, Edmonton

Head Office: Room 1, Burns Block, Calgary, Alta.

The Standard Publishing Co. Ltd., [illegible] Eighth Avenue, W., Calgary.

Despite evidence the Waterton area offered scant oil resources, the Original Discovery Oil Company's 1914 sales pitch for the new venture raised $1 million.

then-staggering $1 million through its public share offering. My reading of the prospectus suggests it was that same million dollars that financed the drilling operation in my photo.

You won't be surprised to hear that Fred and Allan's smooth words proved unfounded. For all the reasons Graeme observed, the Original Discovery Oil Company was a money-losing proposition. Waterton never delivered meaningful quantities of oil. The million dollars frittered away to nothing, and the well was abandoned in 1920. Modern rotary rigs and Turner Valley to the north were where the real money was made.

After immersing myself in the past, I pulled out photos from my own field trips. Fred and Allan's antique operation paled beside my modern images of giant metal rigs adorned with hefty hydraulic machinery.

As I contemplated both how much and how little the industry has changed, I imagined the same scenario playing out a century later...

"NICE CAR." FRED patted the dashboard of Graeme's new electric vehicle. "The ride's so silky smooth and quiet."

"Thanks." Graeme smiled proudly. "These babies are the future of transportation, you know."

Amy leaned forward from the back seat. "Our daughters love riding in it."

"I bet they do," Fred chuckled. "I feel like a kid myself."

"Well, theirs is the generation that will champion this technology," Amy mused. "By the time they're our age, every wheel in the world could be turning thanks to electricity."

Fred nodded slowly, only half listening. "Here's the exit — should take us to the north side of the oilfield. The drill site is just off the highway."

Graeme guided his vehicle onto the exit ramp. "Let me level with you, Fred. Amy and I are reluctant to invest any more in oil and gas."

"Electrification and fuel economy — that's the future," Amy opined.

Fred scratched a bony knob behind his ear, considering her words. "I know. But I promise you'll be impressed with this operation. The equipment may be a bit long in the tooth, but next year, we'll have high-productivity, low-cost pad drilling in place, giving us incredible economies of scale. We'll be cost-competitive with any energy system, old or new. That's a guarantee."

"I'm glad to hear it, Fred, because I get annoyed when people use our money to fund high-cost, antiquated operations on the tail end of change," Graeme snipped.

"And trust me," Amy added, "if Graeme makes a bad investment, he hears about it from me."

"Never marry your business partner," Graeme joked.

"Well, just wait till you're on the floor of the rig, looking up at that mast," Fred persisted. "You'll see."

Growing impatient with Fred's sales pitch, Amy rolled her eyes and continued Graeme's lesson. "Doesn't matter whether it's drilling oil or installing wind turbines or solar panels. You need to lead change and mind your costs, or you're going to lose a lot of money."

AT THOSE WORDS, I snapped out of my daydream.

Reflecting on the fictitious financiers' comments — past and present — I nodded instinctively. The Graeme and Amy types I've met in business echo my own experiences as an investor: overzealous management that relies on waning technology is always a bad bet.

HUBRIS DEFINED

SOCIAL AND
ETHICAL

PREAMBLE

"HUBRIS DEFINED" IS A cautionary tale about arrogance in the energy business.

This story captures the moment that Richard, my imagined publisher of Standard Oil's in-house magazine, first looks up "fuel" in his new Funk & Wagnalls dictionary. To his delight, the definition doesn't mention coal. This simple exclusion amplifies his hubris about the growth of the petroleum business — an attitude manifested in a column found in my October 1914 copy of the *Standard Oil Bulletin.* Coal is relegated to an "also-ran."

Writing "Hubris Defined" was sentimental for me. I started my career at Chevron, formerly known as the Standard Oil Company of California. Naturally, when I had the chance to buy one of the company's old employee magazines, I didn't hesitate. And then of course I had to seek out the 1914 dictionary it referred to as well.

Together, these two publications capture an important moment in energy history: the transition from coal to oil-fired engines in ships, railroads and road transport. On the eve of World War I, the business dynamics of coal-to-oil were accelerated by military conflict, which provided the strategic imperative to adopt petroleum products quickly.

However, more than a century after that Standard Oil writer bragged about displacing coal with their company's oil products, the lumpy black fuel still reigns in the energy mix of most countries.

As I wrote the story, I imagined applying its message to new-age energy publications today. Like my *Standard Oil Bulletin*, too many of them simplistically predict the imminent end of both oil and coal. "Hubris Defined" reminds readers that displacing deeply entrenched energy systems should never be considered easy.

I recognized that **gloating.**

HUBRIS DEFINED

ARROGANCE ISN'T A quality typically associated with a dictionary. But I'm going to tell you a story about how a seemingly tiny tweak in a dictionary caused its user to make a strategic faux pas I often see in the energy business.

THE PUBLISHER OF the *Standard Oil Bulletin* — I'll call him Richard — was in a good mood. Like any word lover, he was keen to explore his new dictionary. Passing his hand over the cover, he admired the latest *Funk & Wagnalls New Standard Dictionary of the English Language*.

"This 1914 edition is a beaut," Richard said to his coworker Bill. "Let's see what it says about our business." He was itching to look up that first word.

Opening the book, Richard thumbed through its crisp pages till he found the section. His right finger ran down the columns as he scanned for the one word that best described his employer's main product. "Let's see here," he muttered under his breath, "fill, finch, fort, fudge... aha — FUEL!"

"What does it say, sir?" Bill asked with a hint of anticipation. As an editor, he knew full well that, in addition to defining a language, a dictionary subtly reflects a society's contemporary attitudes.

Silence for a moment. Then a smile. Richard raised his left index finger above his head and read aloud with animated confidence, "Fuel: Combustible matter used to kindle or sustain fire or produce heat, as oil, wood, etc."

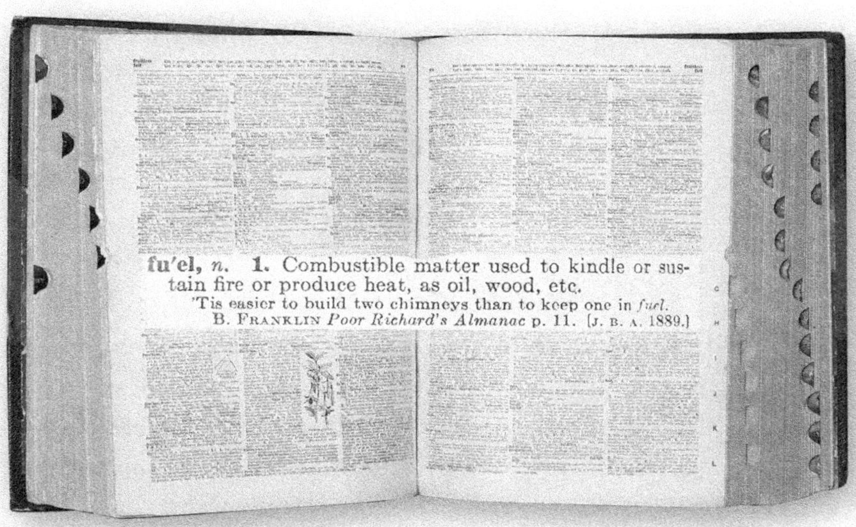

It's no surprise the 1914 *New Standard Dictionary* didn't include coal in its definition of "fuel." By then, coal-fired steam engines were in fast demise, replaced by oil-based engines.

Lowering his arm, he read it again to be doubly sure. He was grinning like the cat that ate the canary when he looked up. "Bill, it doesn't even *mention* coal. It's not in the definition. Ha! Coal is out. Oil is definitively in. Now *that's* a story we can write about!"

Pulling up a chair to his trusty Remington, Richard rolled in a fresh sheet of paper and began plunking on the keys, transcribing the definition.

"PLEASE note," he went on, "and it may be taken as a sign of the times — that oil is set down first of all." Swiping a firm carriage return, he continued. "Coal is not even mentioned; those dictionary-makers have already counted out coal as a fuel, or at best, leave it to the imagination of the reader to include it among the also-rans!" Great introduction, he thought smugly.

Richard was an efficient writer, so it didn't take him long to finish the column. The gist: diesel engines and furnaces were changing the world. In the process, the once-mighty fuel called coal was being displaced by growing supplies of crude oil and refined petroleum products from his own company.

He readied his final sentence with unrestrained passion for Standard Oil and for providing the world with this new type of energy. "The convenience of this service to steamship operators can readily be imagined," he typed, "and the Company has materially added to its fuel oil business because of it."

That done, Richard rolled the last piece of paper out of the typewriter, tidied the pages together and handed over his work. "Bill, let's put this on page four."

The young editor brightened. "Sure thing, boss. Those coal guys are goin' outta business!"

“Coal is not even mentioned; those dictionary-makers have already counted out coal as a fuel…”

The Harm of Hubris

That article, “To Sustain Fire or Produce Heat,” was published in the October 1914 *Standard Oil Bulletin*. Over a hundred years later, in an antiquarian bookstore, I bought a copy.

I started my career at the Chevron Corporation, originally known as the Standard Oil Company of California. That’s one reason I had to add this publication to my collection of energy ephemera.

But the other reason was the lesson embedded in this article. I recognized that gloating.

For many years I’ve studied macro trends, business tactics and investment strategies. What has struck me repeatedly is the suddenness of changing fortunes: how quickly companies, even entire energy industries, can swing from the euphoria of

"FUEL—Combustible matter used to kindle or sustain fire or produce heat, as oil, wood, etc."—New Standard Dictionary.

PLEASE note—and it may be taken as a sign of the times—that oil is set down first of all. Coal is not even mentioned; those dictionary-makers have already counted out coal as a fuel, or at best, leave it to the imagination of the reader to include it among the also-rans!

The *Standard Oil Bulletin* gloated about coal's exclusion from the definition of "fuel," but coal has stayed in the game for centuries, and energy leaders should never count out such low-cost energy systems.

commercial success to the long faces of failure. Personally, and by watching others, I've learned the folly of counting out a competitor prematurely — and of doing so with arrogance.

That's lesson number one from Richard's story: avoid permeating any hint of hubris into company culture.

This isn't to say I don't like positive-minded messages. What makes me uncomfortable about this article is the writer's hasty

I've learned the folly of counting out a competitor prematurely — **and of doing so with arrogance.**

notion that coal was an "also-ran." Sure, he grants that it "might have been cited as a pretty good example of combustible matter still somewhat extensively used for making fire," but that's a lot of qualifiers. His position is clear: coal is a has-been fuel on its way to obsolescence.

Richard and his real-life counterpart joined a long line of naysayers quick to write off coal. In fact, I have books going as far back as the 14th century that predict the demise of coal!

In 1914, their exuberance was understandable. By October, World War I was only three months old. Coal-fired ships and railroad locomotives were quickly converting their engines to petroleum. So were army tanks and vehicles, benefiting from the mass commercialization of engines in cars like the Ford Model T.

Despite the use of oil in planes, trains and automobiles, coal hasn't lost much ground in the last century. After oil, it's the second-most-used fuel in the world, rebutting many attempts at substitution along the way, and sometimes even regaining ground. During the oil crises of the 1970s, coal pushed oil out of the market for electrical power generation.

So why has coal been so resilient for so long?

By any measure, coal's ongoing growth as a cheap, economical fuel for things like power generation has been remarkable. The reason this black lumpy stuff has endured — even despite its negative environmental effects — relates to one of the most important lessons in the energy business: low-cost fuels are highly resilient and difficult to displace.

Both the Funk & Wagnalls editor and the *Bulletin* writer failed to acknowledge the tenacity of an entrenched business that supplies cheap products. Their premature interpretation of world energy markets in 1914 shouldn't be lost on today's leaders, especially those competing to shape our energy future.

Yet as I flip through the *Standard Oil Bulletin*, it strikes me that this same hubris exists heading into the 2020s, as wind, solar and other renewable energy sources challenge the market share of coal, oil and natural gas. Headlines proclaim the end of fossil fuels with Richard's same high-spirited enthusiasm. Implicit in them is echoes of denial of the incumbents' strength. And all too frequently, I've watched companies lose valuable time and money when being cavalier about the aggressive resolve of an entrenched competitor.

My advice to anyone in the energy business, regardless of their product, is to never underestimate the staying power of a low-cost producer that delivers easy energy. If such an urge arises, I suggest grabbing a dictionary and looking up two words: "hubris" and "denial." Commit them to memory but never put them into practice — together, they can take down businesses.

NOBODY TIPS A SCANDISCOPE

SOCIAL AND ETHICAL

PREAMBLE

When I started writing "Nobody Tips a Scandiscope," I was pretty sure the central lesson was going to be about the ethics of energy. Behind the comforts and benefits of the light switch, the gas pump or, in this case, the fireplace, there's often upstream human sacrifice.

This story's ethical message is clear. It's painfully amplified whenever I go to European cities like London. Looking at those tall chimneys, I think about the enslaved children who were forced to crawl through narrow brick flues caked in the toxic soot of 19th-century coal combustion. What agony did those children suffer? What kind of people sanctioned such a horrific practice?

But as I read more from my 1825 book *The Chimney-Sweeper's Friend, and Climbing-Boy's Album*, I realized there was a deeper question: Why did child chimney sweeping continue for a hundred years after the invention of technology that should have snuffed out the practice?

The answer lies in the selfish interests of individuals involved in the economy of chimney sweeping. Both chimney masters and the housekeepers of the homes they visited had a stake in prolonging the practice, making unfamiliar technology like the Scandiscope a tough sell.

These days, “Nobody Tips a Scandiscope” sits in the back of my mind as I think about the barriers to adopting products and processes that reduce environmental impacts of energy use. I don’t look to technology as the barrier, I look to the economics of human behavior.

I hope this story resonates with you, as it did with me. And the next time you’re in Europe, look up at the chimneys. You’ll shake your head at the practice of using children as chimney sweeps. Shake your head too at why it took so long to save them.

Why did society **push innocent children up chimneys** for so long?

NOBODY TIPS A SCANDISCOPE

I climb'd, and climb'd, and climb'd in vain,
No light at top appear'd;
No end to darkness, toil, and pain,
While worse and worse I fear'd.

"The Climbing-Boy's Soliloquies," *The Chimney-Sweeper's Friend, and Climbing-Boy's Album*

THE SIMPLEST THINGS can sometimes teach us the most. That's why I want to tell you a story about chimneys.

A chimney is simple enough: it's a hollow column made of bricks, stone or cement that vents out gases from the fuels we burn. But within those flues lurk eye-opening lessons about the influences on, and implications of, our energy choices.

To get in the mood for this story, do what I'm doing as I write it: settle into a comfy chair by a crackling fire late at night. Absorb the heat and let your mind wander. Imagine you're in England's north country, where it's damp and cold. Now imagine it's January 1857.

HENRY HAWORTH was nervous.

He was standing at the gates of a 16th-century school, looking up at the roof of a large three-story stone building. The glint of the early morning moon backlit four cylindrical pots, grouped in a row on top of the chimney.

Only seven years old, Henry already knew his chimneys. He knew them inside and out. Literally. You see, Henry was a chimney sweep, although that was a benign description of what he did for a living.

In mid-19th-century Britain, most people would have referred to Henry as a climbing boy, which in my mind was a more fitting handle. That's because he didn't really sweep much. Instead, he spent his long days climbing inside fireplace flues, removing layers of soot with his head, hands and a scruffy straw brush.

Before the end of this January work shift, Henry would have to squeeze through each of those four suffocating chimneys.

Like any child, he'd learned quickly from experience: the first stack usually vented the main-floor fireplace — he dreaded that one the most. Filthy stone flues that snaked through walls for 20 meters were common. Add to that 90-degree kinks jammed with dregs still hot from the previous night's fire, and a flue seemed to go on forever, all the way up to the clay pot's narrow opening at the end of the stack.

Henry kept staring at the top of the building.

Having never gone to school, he could neither read nor write. But he could add up what scared him. Four, he thought, four of the scariest tall-pot chimneys. All to be cleaned before noon!

The boy's worrying was interrupted by his foul-tempered chimney master.

"Stop starin' and git a move on!" barked Nigel. "We got a lot to do this morning." His snarl revealed a mouth full of rotten teeth.

Henry took a deep breath of crisp winter air. It would be his last pleasure of the day. Head hanging, he limped behind Nigel through the school gates. No one cared that his left ankle was sporting a bloody scrape, inflicted by one of yesterday's merciless chimneys.

Might to Blight

Most people don't notice chimneys, but I do. As an energy tourist, I take pictures of them, especially when I travel to old European cities with rich crops of chimneys sprouting out of urban landscapes. Why? Because chimneys have been fundamental to the evolution of our energy needs.

By the 15th century, chimneys were gaining popularity in Europe as a way to separate smoke from the warmth of an indoor fireplace. Soon after, steam engines started pulling, pushing and turning, and it wasn't long before mechanization led to electric power generation. All these fuel-hungry contraptions needed a stack for blowing out the energy we *didn't* use in the process of creating economic prosperity and comfort. With the Industrial Revolution, chimneys came to signify productivity and industrial might. Now, they symbolize two of society's great energy ills: inefficiency and environmental degradation.

I admit I hadn't given chimney sweeping much thought until I bought a rare 1825 book titled *The Chimney-Sweeper's Friend, and Climbing-Boy's Album*, edited by James Montgomery. By 1825, the practice had already been in play for over 150 years, and Montgomery, a poet, newspaper publisher and tireless advocate for the helpless children, documented much of the horrific history. Intrigued and appalled by what I read, I looked for

Once you've heard the stories of the climbing children, you never look at chimneys the same way.

more materials on the climbing boys' plight. My hunt yielded "Climbing Boys," an 1857 article by an unknown author. It opens with a quote that's ominously on point: "They die in their youth, and their life is among the unclean."

Between this short essay and Montgomery's 500-page treatise, I'd added to my library two of the most depressing pieces of literature I own. Their pages are filled with heartbreaking stories of climbing boys being burned, beaten or suffocated to death. Like John Anderson. At five years old, he'd already been working for a year and a half. One day, he was sent up a flue still hot from the night before. Within five minutes, people could hear him groaning, but they couldn't reach him in time and the poor boy suffocated.

And so, I added ethics to the list of energy ills symbolized by a chimney.

HENRY WAS THE middle child of three, born to an impoverished family that desperately needed money to survive. Six months earlier, his mother had made the unimaginable decision to sell her son, a perfectly aged and physically able boy, into Nigel's apprenticeship.

"Here's three shillings fer ya," Nigel had sneered, dropping coins into her shaking hands.

"Come here, boy." The sweep grabbed Henry's arm, claiming his newly acquired asset.

"Please, Mum, please! Don't let him take me away," Henry had whimpered, straining his free arm out to his mother. But it was too late. The deal was done, and she was already retreating into the shadows of a narrow street.

With that, Nigel became Henry's chimney master or, more bluntly, his slave master.

Signing Their Lives Away with an X

Not all climbing boys were sold into slavery by their parents. Many in England were orphans offered up by workhouses, state-run dwellings for the destitute. In fact, chimney masters were often paid to take children away — for as little as $20 in today's terms — to ease the state's burden of caring for the neglected.

During the transaction, a child would have to sign an indenture, usually with an X, in front of witnesses, pledging seven years to his new vocation. In return, the boy's master was responsible for feeding, clothing and sheltering his new charge.

Many good masters fulfilled their end of the bargain, but it's well documented that plenty of unscrupulous ones fell short of delivering on their basic responsibilities. In fact, undernourished boys were desirable because they were better able to climb narrow passages. How narrow? Open a magazine. That 11-by-17-inch spread is larger than the standard flue child workers had to squeeze through.

Thinking about such conditions reminds me of *Oliver Twist*, in which Mr. Gamfield, a master sweep, discusses how stuffing burning straw up a flue motivates climbing boys: "... even if they've stuck in the chimbley, roasting their feet makes 'em struggle to hextricate theirselves." The chair-squirming truth is that Dickens' fiction is based on facts openly published at the time.

Girls weren't exempt either. Some were pressed into service if their chimney master parents didn't have sons to support the business. Their slighter frames also made them better able to fit into those tiny flues.

NIGEL LED HENRY to the fireplace, still oblivious to the young one's limp. Before the boy got started, he took off his shoes and brought out his climbing cap. He pulled the improvised

Chimney masters were often paid to take children away — for as little as $20 in today's terms.

balaclava over his head and below his chin, to prevent ingesting the toxic residue.

Henry hated this part — preparing for the dark unknown. He wished his mother was by his side, to kiss his ankle better, to soften his fears. But she wasn't. Hoping to rid himself of memories of her, he gave his head a shake and grabbed a small straw brush. He launched himself up off the hearth's grate and pushed into the impossibly tight flue by pressing the sides with his back, elbows and knees, articulating his body like a caterpillar.

Scraping, brushing and squirming, Henry loosened soft and hard soot for the first nine meters, allowing the residue to freely fall into the fireplace. Later, Nigel would put the pitch-black soot in sacks and take it to the afternoon market, where he sold it to farmers as fertilizer.

Henry kept plodding skyward until he poked his head up to a horizontal stretch. Pulling himself around the corner on to the ledge, he found momentary relief from the force of gravity. But now he had to scrape forward, channel the soot underneath his belly, then push it back with his feet. This repetitive forward-reverse motion gradually thrust all the residue to the free-fall of the vertical segment he'd just cleared.

After a few meters, Henry's head touched a wall. Catching his breath, he sensed open space above. From experience, he knew he was about to bend upward into the final vertical section. Relief set in. Soon he would be done this first flue.

The Long Road to Saving the Children

While the energy geek in me has always been interested in European chimneys, I look at them differently now. Examining

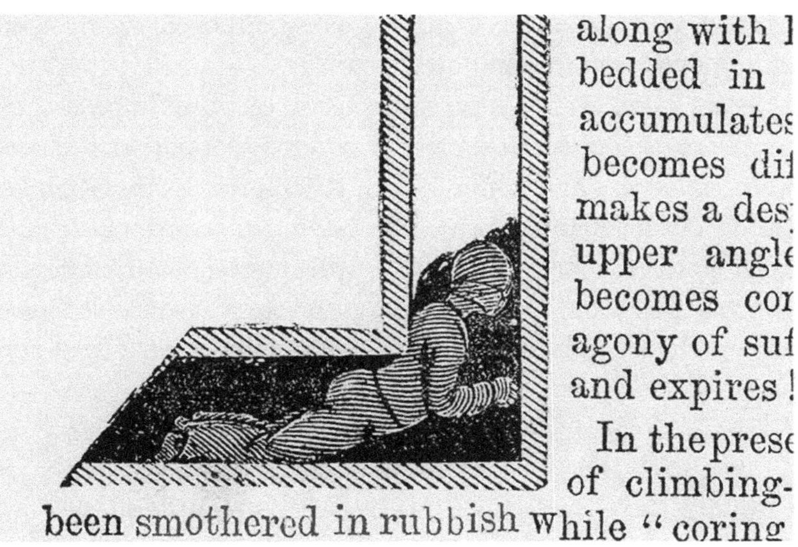

Suffocation was the most common cause of death for climbing boys. Squeezed in a flue no larger than 11 by 17 inches, a child had no hope of escape should soot collapse, dying in "an agony of suffocation."

them inside and out, I imagine those poor climbing children and reflect on what their horrors mean to us today.

The obvious takeaway is that we must be mindful of the ethical issues associated with our energy systems. I'm not naive to what goes on behind the scenes of our Western comforts. Nor should you be. The gasoline you put in your tank may have originated from an authoritarian state that tortures its citizens. The battery in your phone, weed whacker or electric car may contain scarce metals from Africa, chipped out of a mine by a child. These are 21st-century realities.

But as I sit in front of my fireplace, I'm thinking about a bigger issue. The first ban on the use of climbing boys came in 1788. Mechanical chimney-sweeping devices were around as early as 1789. Why did society continue to push innocent children up chimneys for another hundred years?

From a callous business perspective, cost and implementation were two big reasons. Early chimney-sweeping innovations were expensive and complicated. It took until the 1870s for the cost of mechanical brushing systems to come down. Many master sweeps couldn't afford the upfront cost of a mechanical contraption. If they could, they argued it took too long to figure out how to use the device properly, especially in hard-to-sweep areas like horizontal sections.

So it was cheaper and more effective to use youngsters.

In 1803, the Society for Superseding the Necessity of Climbing Boys was founded. Dedicated to saving climbing boys from slavery, the organization also promoted the use of mechanical technologies. Encouraged by the society's formation, the Royal Society of Arts revived its languishing design competition for chimney-sweeping machines. In 1805, George Smart's "Scandiscope" won. The Scandiscope was a series of hollow rods held together by a long cord that ran down the inside. The chimney sweep connected the rods one by one, creating a long-handled straw brush he pushed up the flue, scrubbing the chimney.

Other innovations followed. In 1828 Joseph Glass introduced a similar kit that used screw-together brass rods to extend the long handle. Considered superior to the Scandiscope, Glass's tool began penetrating the sweeping market in earnest, but there were still many years to go before climbing boys were eliminated.

These innovations were paralleled by legislation that went largely ignored. Adherence to the 1788 ban was minimal, especially in rural England and Wales, where child laborers were cheapest. Authorities notionally accepted the master sweeps'

Because technology can flatten hierarchy, it isn't always desirable.

arguments against the chimney-sweeping contraptions, turning a blind eye to the legislation.

The rapidly increasing use of coal during the Industrial Revolution was another compelling reason to let the practice go unchecked. Chimneys were being built at a frenzied pace, and sweeping was an essential service for fueling prosperity.

But basic economics wasn't the primary headwind against the adoption of mechanical cleaning. As I read more about climbing boys, I realized that the persistence of entrenched interests was the greatest source of friction — the slow pace of change wasn't due to a lack of technology, it was thanks to selfish, human fixations.

Take social status, for example. In many 19th-century cultures, the privileged class didn't have much interest in changing its standing in life. Society's bottom rung, represented by individuals like climbing boys, was necessary to the preservation of class distinctions.

In thinking about this, I pause to look at the illustrations in Montgomery's book: top-hatted gentry point at and disparage

The social strata of the sweeping economy collide: the upper class looks on as a chimney master acquires a new apprentice. Resistance to change has many sources. In the case of climbing children, maintaining class distinction was one.

climbing boys. Social status is important to egos, and because technology can flatten hierarchy, it isn't always desirable.

When analyzing the potential for change, I think about the phrase "Money talks." Recognizing entrenched financial interests means knowing where the talking, or whispering, is going on.

In the business of climbing boys, there was a small underground economy of perquisites that were massive barriers to change. Housekeepers, the primary decision makers about cleaning chimneys, would receive tips and perks when their homeowner's chimney was kept in good order. If a chimney master sullied carpets and furniture with a new contraption, no tips were forthcoming. So their resistance was financially motivated, for fear their perks would be jeopardized by this disruptive technology.

Similarly, master sweeps had perks at stake. Servants and homeowners would often sympathize with pathetic-looking climbing boys, giving them tips and food at the end of a job. Both dividends inevitably ended up in the master's hands. Bottom line: nobody tipped a Scandiscope, a fact every master took into the calculus of adopting new technology — or not.

> Some wooden tubes, a brush, and rope
> Are all you need employ;
> Pray order, maids, the Scandiscope,
> And not the climbing boy.
>
> *The Every-Day Book; Or, Everlasting Calendar of Popular Amusements, Sports, Pastimes, Ceremonies, Manners, Customs, and Events, Incident to Each of the Three Hundred and Sixty-Five Days, in Past and Present Times*

HENRY BEGAN CONTORTING his body upward into the next vertical section, but he still couldn't see any light above him. Relief turned to worry. No light almost always meant a blockage.

He was right.

Suddenly, a rush of soot collapsed, enveloping his head and shoulders.

Panic overtook the child. He couldn't move. Breathing was difficult. "Help!" he gasped. But it was futile — no one could hear him.

Within minutes, this seven-year-old suffocated to death, his body entombed as a wasted life, cleaning waste from an appliance that emitted wasted energy.

THE EVERY-DAY BOOK.—MAY 1.

Northampton May Garland.

617 THE EVERY-DAY BOOK.—MAY 1. 618

The last Chimney Sweeper.

A large brush made of a number of small whalebone sticks, fastened into a round ball of wood, and extending in most cases to a diameter of two feet, is thrust up the chimney by means of hollow cylinders or tubes, fitting into one another like the joints of a fishing rod, with a long cord running through them; it is worked up and down, as each fresh joint is added, until it reaches the chimney pot; it is then shortened joint by joint, and on each joint being removed, is in like manner worked up and down in its descent; and thus you have your chimney swept perfectly clean by this machine, which is called a Scandiscope.

Some wooden tubes, a brush, and rope,
Are all you need employ;
Pray order, maids, the Scandiscope,
And not the climbing boy.

Copy of a printed hand-bill, distributed before May-day, 1826.

No May Day Sweeps.

CAUTION.

The inhabitants of this parish are most respectfully informed, that the UNITED SOCIETY OF MASTER CHIMNEY SWEEPERS intend giving their apprentices a dinner, at the Eyre Arms,

Although this May 1826 handbill dubbed the Scandiscope “the last Chimney Sweeper” and exhorted housekeepers to use the new tool, it was another 50 years before people ceased sending children up chimneys.

A Force to Be Reckoned With

The crackling fire beside me underscores the silence in my mind.

Think about it: it took five parliamentary edicts and a *hundred* years before the Chimney Sweepers Act of 1875 ended the era of climbing boys for good.

Looking up at the stone stack, I wonder how to end this story.

The obvious conclusion is that child labor is despicable. Or that unethical practices in the supply and use of energy are ill-advised. Both are vital lessons, but the primary business lesson lies in the troubling longevity of climbing boys: deeply rooted institutions in our society don't change as fast as technology.

A century and a half later, we're still trying to displace chimneys, stacks and exhaust pipes — the conduits of combusting fossil fuels. This despite the fact that more efficient energy processes are readily available. Climbing boys are a testament to the difficulty of uprooting entrenched interests and low-cost processes. When thinking about our energy future, across all energy systems, do we understand those ingrained interests?

Reflecting on my own experiences in the business world, I realize we talk far too much about the potential of new technologies — how new product A will clobber old-school product B — and too little about matters related to people, especially the hardiness of entrenched interests at a raw, human level.

In fact, technology and "know-how" are rarely the limiting factors in a successful business plan, or in making the world a better place. This tale of Henry Haworth and his climbing peers should remind us that self-interest can be so rigid, so strong, that it can seriously hinder the adoption of superior solutions — and abet even the worst human practices.

THE GREAT AHA

SOCIAL AND
ETHICAL

PREAMBLE

DENIAL. IT'S THE REFLEX of rejecting reality when all the trends, all the signals, all the scuttlebutt points to the inevitability of change. I see it often in the business world. And I confess that, on occasion, even I've found myself in denial when the forces of change were barking at my heels. We're all human.

That's why the story about Ed Tucker in a 1914 issue of *Ford Times* resonated with me. The owner of a successful stagecoach operation, Ed was in denial about the automobile taking over his livelihood — till he had his aha moment.

"The Great Aha" recounts Ed's decision to shutter the stables and buy a car, a Model T to be exact. On the verge of making a major investment decision to expand his business, this plain-spoken pragmatist realized that more of the same in the face of change would allow competitors to take over his reins.

But this story is about more than one man's business choices. It shows why Henry Ford's iconic car was an instant success: it was a compelling, direct substitute for the horse and buggy, and at lower cost. No wonder its adoption was so rapid and widespread.

Extreme cases of substitution like this are easy to figure out, but not every transition is as straightforward. Are today's electric cars as compelling to consumers as a Ford Model T was a hundred years ago? The answer lies not so much in the positive attributes of a battery-powered car, but in consumers' resistance to deviating from their comfort zone.

So, in a way, "The Great Aha" is an exploration of the opposite of denial: exuberance. In an era when robots, artificial intelligence, green energy, flying and autonomous cars are driving headlines, Ed Tucker's story helps mediate reality between the extremes of denial and exuberance.

What would have happened if the Model T had only been as good as, or marginally better than, a team of horses pulling a quality carriage?

THE GREAT AHA

EVERYONE IN THE village of Kalkerville knew the Tuckers. Young Ed operated the local stagecoach business, moving the locals, their goods and their mail. For years, Ed's dad, Jim, had been the familiar face sitting on top of the coach, wheedling the horses. But after the elder died, Ed picked up the reins of the family business.

Like his father, Ed was a no-nonsense skeptic who didn't have a lot of patience. Quick to offer advice, and slow to take any, Ed had a tendency to shut his mind like a barn door.

By early 1914, the occasional motorcar could be seen puttering down the streets of Kalkerville, piquing curiosity among locals. Not Ed. It didn't take more than one drink to put a smirk on his face and get him ranting: "None of them automobiles could pull a load of folks and trunks through the deep sand," he'd say loudly to friends, "and do it day after day."

Ed knew the robust reliability of a stagecoach pulled by trusty horses. Rain, snow or shine, his team of two, sometimes four, animals could haul paying passengers from Kalkerville over the big hill to Dart — a 23-kilometer journey — in half a day. Deadheading wasn't a problem: on arrival, six returning fares would be waiting at the Dart station, near the saloon on Main Street.

Most days, Ed was home before 8:00. He'd sit down for a late meal with his family, tell tall tales about the day, then tend to the horses and get ready to do it all over again the next morning.

Ed appeared to lead a simple life, but running a stagecoach line wasn't easy. Repairing the canopy-covered coach was a relentless job. The wooden wagon wheels took a beating and needed repair after almost every trip.

Harnesses, yokes and hinges were prone to damage, especially after being pulled in bad weather — mud and snow were particularly unforgiving. Of course, the greatest obligation wasn't mechanical, it was the time and expense of caring for the horses.

Despite these responsibilities, as far as Ed was concerned, there was no way a noisy mechanical contraption called a "horseless carriage" could compete with the tried-and-true stamina of animal power. "After all," he was fond of saying, "horses and carts have been moving folks around since the time of them Romans."

Those Nags'll Eat Your Profits

I came across Ed's story in the Canadian edition of *Ford Times* from April 1914. As usual, I found this nostalgic gem in a disorganized antique store, under a pile of neglected *Life* magazines, superhero comics and tattered calendars.

Ford Times shilled the virtues of owning a Ford automobile. Back then, the rage was the now-iconic Model T, including the touring car, runabout, town car and coupe. Blurbs like

"Play it safe — buy a Ford" or "Less than a two-cent stamp is the cost per mile of Ford travel" are splashed throughout the 45-page issue.

"Knowing the Ford," Ed's story, is toward the end. Before I started reading it, two cartoons caught my attention. One shows a pair of horses gobbling up dollar bills from a box labeled "PROFITS." In case the message isn't obvious, the caption spells it out: "Where the profit used to go." Businesses that relied on horses saw their profits eaten up — and were already in the past tense.

Flipping the page, I chuckled at the second anti-horse cartoon. A couple of unresponsive nags pull a dilapidated stagecoach. Its smug caption states unequivocally, "What the Ford replaced."

Where the profit used to go **What the Ford replaced**

In the early days of the Model T, *Ford Times* liked to belittle the entrenched competition, a hallmark of corporate house organ publications.

Automobiles quickly liberated horses from their monotonous pulling chores.

That word — "replaced" — fast-forwarded my mind to today, when governments and businesses want to accelerate the transition away from or, more bluntly, replace the wasteful, polluting elements of our legacy energy systems. For example, there's a push to replace billions of inefficient yet deeply entrenched devices, like incandescent light bulbs with LEDs, or old air conditioners with new energy-efficient ones. Furnaces, refrigerators and anything that recklessly guzzles fuel is a target. More topical — and analogous here — is all the excitement about swapping out petroleum-powered cars with the next era of personal transport: electric vehicles.

We know how events unfolded in the early 20th century: automobiles like the Ford Model T quickly liberated horses from their monotonous pulling chores. With that, carriage makers went out of business, and hardheaded stagecoach operators succumbed to the same disruptive forces of mechanization.

But the benefits of that transition were tangible. The challenge today — and it's a big one — is to convince skeptical individuals that the contrast between what they already own and its replacement is as stark as the difference between a rickety stagecoach and a shiny new car.

Clarity in a Cloud of Dust

Ed was clearly in denial of change, staid in the same views inherited from his father's meat-and-potatoes DNA. Yet his aha moment came one day as he "was plodding along with a load toward Dart."

I imagined Ed daydreaming while sitting atop his coach full of six passengers. The creaks and jangles of his rig were hypnotic as he stared mindlessly at the same old ruts in the road.

Suddenly, an unfamiliar noise broke the monotony. "A Ford touring car filled to the brim with young people — six, to be exact — pulled out from behind and went whizzing past."

I love how the author remarks on the passengers' youth. Even back then, progressive-minded 20-somethings were the target market for new technology.

Seeing the automobile "go spinning up the road until it went out of sight in a cloud of dust" caused Ed to think pretty hard. The passenger load between Kalkerville and Dart was getting busy enough to justify investing in another stagecoach and more horses. Ed had been wondering if he should make the investment. Sure, more potential to make money was there. But the additional costs of keeping horses and hiring people would dig into profits.

Lucky for Ed, the Ford that whizzed by had him thinking all the way to Dart. It was that three-and-a-half-hour ride that invoked his "Aha!"

The rest of the story unfolds as you'd expect. Ed bought a Model T touring car and started hauling mail bags and people.

“I can go back a’whistlin’,” he boasted. “Made the fourteen miles in just fifty minutes. Clipped about two hours and a half off the team’s time without half tryin’.”

Three years later, Ed’s Ford was a Kalkerville institution, respected and admired by the whole community. His customers were happy because they had faster transport in greater comfort. Ed was happy because his per-trip profits were higher.

In fact, Ed’s fortunes were even better than he’d imagined. Instead of one round trip, he could make as many as four, quadrupling his already fatter bank account. Pulling ability also played to his favor: his strongest horse, Jimmy, could pull almost as much as a Model T could, but only for a moment. By contrast, Ed’s new car could be seen “whizzing up the sand road as if it hadn’t any load at all.” Finally, Ed had more time in the evening because he wasn’t working on his carriage. His Model T had been performing well for three years and was “good for as many years more.”

All this sounded great, but the story makes no mention of price, so my usual curiosity led me to do some research. In 1914, a new Model T touring car could be had for $550. By contrast, a premium horse-drawn McLaughlin carriage had a sticker price of $150, but that didn’t include all the gear, such as harnesses, reins and yoke. Nor did it include the horses and barn to keep them in.

So, in my estimation, the cost to set up a transport company in 1914 was nearly the same, whether with a stagecoach or automobile — about $550. In fact, all expenses considered, the Model T might even have been cheaper.

Same price up front. Four times the productivity of a team of horses. Five times the power. Longer range. Greater comfort. More reliability. Lower operating costs and maintenance. This was the proverbial no-brainer.

Adoption of the Model T was swift — within a decade, it had penetrated markets worldwide. More than just a mass-produced product, the "Tin Lizzy" was a society changer.

From Ho-hum to Aha!

At the risk of showing my age, I remember being taught how to use a slide rule in the last year of high school. A classmate boldly asked our physics teacher, "Why do we need to learn this? Everyone is starting to buy calculators."

Our teacher shrugged and smiled, tacitly admitting the looming obsolescence of the two sliding sticks engraved with tick marks. But it was part of the laggard school curriculum, so we had to go through the motions of learning how to manipulate the same tool Isaac Newton had used when he was a budding math nerd in the 17th century.

Being a geek myself, that same year I used my savings to buy my first Texas Instruments electronic calculator, validating the step change in solving math problems.

That was the beginning of many similar leaps I've made between what I call "high-contrast" technological innovations. In other words, the difference between inferior old stuff and new stuff that's irresistibly compelling: CDs over vinyl, computer over typewriter, flat-screen monitor over tube TV, GPS over paper maps, smartphone over dumb phone and streaming movies over DVDs.

Then there's the stuff that wasn't all that much better from what was in use. The laser disc, 3D television, satellite phone, Apple Newton and a long list of doomed products — new technologies that didn't make it, in large part because they weren't different enough from what people were already using. Quite often they were more expensive too.

As I finished "Knowing the Ford," I wondered what would have happened if the Model T had only been as good as, or marginally better than, a team of horses pulling a quality carriage.

Imagining myself sitting in the Kalkerville saloon, I could hear Ed's boisterous three-whiskey voice blurting out, "Why would I buy one of them things? I bet a round of drinks that my two best horses could win a race to Dart and back! For less money too!"

And he might have been right.

If people don't see the need, a product **runs the risk of going the way of so many other ho-hum products.**

Had the benefits of the Ford not far outstripped that of the reigning mode of transport, adoption of the motorized vehicle might have plodded along at the pace of a rickety old stagecoach.

And what of electric vehicles? In the last century, their appeal has gone through fits and starts, but guppy-looking vehicles like GM's EV1 were never able to deliver that no-brainer utility to car buyers.

Yet, almost a hundred years after my issue of *Ford Times* was published, a new era dawned. In 2012, Tesla introduced its sexy Model S sedan. Soon after, its whiplash-inducing performance won it the coveted *Motor Trend* Car of the Year. Then came the Model X SUV. Both the S and the X did well, but at $100,000-plus for a set of wheels, they were targeted to the affluent.

With the 2017 launch of the Model 3, Tesla finally offered a more affordable model built for the masses. Many believed it was the turning point for electric vehicles, much as Ford's Model T was a century prior. GM and Volkswagen quickly jumped into the fray, while Chinese companies BYD Auto and Geely jockeyed for position in the world's largest consumer market. It was easy to understand the excitement. But I was still skeptical about the pace of change.

I'm telling Ed's story in the early days of the electric vehicle revolution. New-age automakers have yet to win over the pre–aha Ed Tucker types, those unyielding naysayers who are perfectly satisfied driving their gas-guzzlers. While the need to eradicate carbon emissions spewing out of a billion exhaust pipes is obvious, it's clearly not an "Aha!" for the majority of consumers. In fact, North Americans are uprating their personal transport needs to even bigger pickup trucks and SUVs.

A large part of my career has been based on assessing the likelihood of nascent technologies displacing what's deeply rooted

Unless a technology can compete against a new innovation on cost, quality and — often most importantly — how well it satisfies a human need, it will suffer the same fate as the carriage: becoming a curiosity from days gone by.

in society. Often I'm not simply deriving conclusions from a spreadsheet — I consider the human side of the equation too. Does this technology address a real need? If people don't see the need, it runs the risk of going the way of so many other ho-hum products.

After many reads of my *Ford Times*, I've realized Ed's story is about more than the evolution of personal transport. It's about human need, and how it informs the broader question: "What makes people buy new, disruptive products?" When it comes to pitching out the DVD collection for Netflix, the answer is obvious. But not all innovations are such slam dunks.

So, if in doubt about the business case for a new product, I always ask myself, "Is *this* an Ed Tucker moment?"

STAIRWAY TO HELL

ECONOMY AND
BUSINESS

PREAMBLE

POPULAR BUSINESS CULTURE IS awash with success stories. Tech entrepreneurs who go from zero to billions are feted in books and magazines and across every media channel. Their accomplishments are impressive and often inspiring.

But what about failure stories? We don't hear them quite as often. That's because demise is a lot harder to acknowledge and discuss. But, as you and I both know, much can be learned from failure. Some argue that falling down a few times is a prerequisite for success. I tend to agree.

"Stairway to Hell" is about corporate failure in the extreme. So extreme that the 1922 insolvency of a local coal mining company wiped out Bankhead, a once-vibrant community of 1,500 people in the Canadian Rockies.

I've been to Bankhead at least 10 times, wandering through its deserted remains. As I've explored the ghost town, I've relived the spirit of its people, its history and its tragedies — and I've learned a lot from its downfall. What's eerie is how much this story can be juxtaposed with corporate tales from the energy business today.

While Mine No. 80 was plagued by many issues, including low-quality product and high-cost production, "Stairway to Hell" boils down to one tenet: hope is not a strategy. Cliché, yes, but, in the slipstream of failing enterprises, it's a pervasive dictum.

Strangely, the tale of Bankhead's demise is more inspiring to me than a typical success story. Its insights into what causes failure can be flipped around to reveal lessons for success for us all.

“**What the hell** happened here?”

STAIRWAY - TO HELL -

JASON AND KATIE are in love. It says so on the crumbling concrete wall in front of me. Or at least they were in love, a long time ago, when Jason took a pencil and firmly inscribed his name above Katie's inside a big heart.

But this story isn't about love. It's about economic misery. It's about how quickly an entire town of workers, families, children and young lovers can descend into a pit of nothingness. It's about decrepit homes, factories filled with rusting equipment and what's left of a church on a hill.

No Signs of Life

I'm walking around Bankhead, at the base of Cascade Mountain in the Canadian Rockies. Hidden behind trees in a picturesque valley, Bankhead is an old coal mining village at latitude 51.231550 and longitude -115.522143 that was once home to

1,500 people. It's good to know the coordinates because there are no streets left in Bankhead. The only sign that this ghost town exists is a small brown one on a secondary Alberta highway.

However, Bankhead isn't completely lost to the world. Most days, a few curious sightseers come from the nearby tourist beehive of Banff. On the road to Lake Minnewanka, they catch sight of that brown sign, turn into a parking lot and maybe have lunch at one of the lonely-looking picnic tables. From there, a short hiking trail loops a kilometer or two around what used to be the town's mining operation. Plaques on squat guideposts inform visitors of the local history and point out what remains of the landmarks on the multi-acre site.

I'm standing in one of Bankhead's empty buildings. The dirt floor is littered with pine cones, one of which I inadvertently step on. I look up and get the impression the sky doesn't like what I just did. Bad-tempered clouds spit drops of rain on me through the long-gone roof. But I'm not bothered. This isn't a happy place. In fact, bad weather seems apropos to the mood around here.

Across the room, a branch pokes through an empty window frame. I wonder, Which had less pity for this economic tragedy: business or nature?

There isn't much doubt in my mind. I know some of the history around here. Business was brutal in 1922, the year everything crashed to an end for Bankhead. Back then, cigar-smoking tycoons packed more punch than anything nature could throw at this town. A boardroom decision to shutter the mine was more devastating than any winter storm passing through the place. Looking around at Bankhead's ghostly remnants, I come to an uncomfortable conclusion: resource-based communities live under the constant threat of oblivion.

The lamp house was once the gateway to the economic heart of Bankhead. Now, the crumbling walls that remain are covered with graffiti.

History's Slag Heap

As someone who studies the energy business, I'm alert to large-scale obsolescence. More than 85% of the world's insatiable needs are extracted from the earth's crust in the form of coal, oil, natural gas and uranium — and plenty of people on six out of seven continents make a living bringing that stuff out of the ground. Now the pressures of change threaten many resource-dependent towns, regions, even entire countries. Their future viability is always hostage to environmental pressures as well as to the dragon that eventually burns every industry in the ass: technological change.

My research into energy transitions, along with field trips to places like Bankhead, has taught me that relegating a business to history's slag heap can happen faster than a bucket of water can soak a lump of burning coal. One day everyone has a job, the next a thousand people are walking the streets in a daze with nothing to do. Years later, mischievous young lovers scrawl graffiti on what used to be the heart of a community's economic well-being, and inside of a century, a curious scribe like me wanders through the rubble asking, "What the hell happened here?"

Do the remains of Bankhead presage the fate of today's oil-producing centers?

In this new age of renewable energy, the parallels are easy to draw. Migration to sources like wind and solar on the front end of energy supply — combined with new alternative devices like electric cars on the consumption end — potentially threatens the livelihoods of millions of people around the world involved in the production of earthly fuels.

It's not just me pretending to be Socrates, philosophizing on a dreary, ash-colored day. Entire countries are thinking about the threat of irrelevance — places like Saudi Arabia, the world's second-largest oil producer. In 2014, policy makers in the desert kingdom had an ontological moment. Recognizing excessive dependency on their potentially limited fuel source, they began thinking about how to diversify away from their oil-drenched economy.

This old mining town is an acute case study that should make **anyone think twice about being complacent about their lot in life.**

Okay, compared to Saudi Arabia, Bankhead looks insignificant, a kingdom only in the minds of those who once lived where I now stand. But scale doesn't matter. Looking around at these vestiges, I think this old mining town is an acute case study that should make anyone think twice about being complacent about their lot in life. Hundreds of communities today could similarly end up as historical footnotes if some tech-savvy alchemist were to figure out how to power a car with Captain James T. Kirk's dilithium crystals instead of John D. Rockefeller's black gold.

A High Price to Pay

I step through the rectangular void where the building's door used to be, hoisting my leg over the crumbling footings. To be sure, Bankhead is an extreme case of desuetude catalyzed by complacency amidst a poor market for coal. But outside the building, I remind myself that jumping to hasty conclusions is as easy as grinning for a selfie these days. I tend to avoid both.

Posing beside rubble and saying "Bankhead went bankrupt, therefore so will all resource towns" is tempting, but the perplexing thing is that the world didn't stop using coal in 1922. Far from it. Today, 7 billion people gorge on five times as much of the black stuff as a hundred years ago. In fact, Bankhead is not a case of a primary energy source becoming obsolete.

I've read a lot about Bankhead, but I still look to the rubble for practical answers to why Mine No. 80 and the town around it went bust. My attention turns to the historical plaque on the outside wall. Apparently the building I was just in was the miners' lamp house, the gathering place for a hard day of back-breaking, thankless work. Closing my eyes, I time-travel a hundred years in the past, to an early April morning in Bankhead.

AT 6:00 A.M., a siren blows, waking the entire community. By 7:45, men are heading to the lamp house for their 8:00 shift. They line up in front of where I'm standing, maybe cracking jokes, shoving each other playfully and cussing about losing at last night's poker game.

Amid the banter, Frank Yakubiec and Mike Perrotti wait to pick up a clean, fully fueled brass safety lamp before they go into the mineshaft. When their turn comes, Frank and Mike get these lamps in exchange for their personal tags, each inscribed with a number.

The two know what this daily tag-for-lamp trade means. They're swapping the valley's natural glories for eight hours of hell inside the bowels of this mountain, bashing and hauling out coal, risking their lives in the process. But they do it, day in, day out — to put food on their families' tables, clothes on their backs.

Frank makes the swap like he does every day. Except on this day — April 22, 1920 — he doesn't make it out.

Records from the Old Calgary Court House diarize that Mike Perrotti "heard coal running down the chute, and about the same instant heard someone shout from inside the chute down below him." That someone was Frank, who "shouted for him to stop the coal from running" and "ordered Mike Perrotti to get an axe and cut a hole in chute..."

It was too late. At 10:45 a.m., Frank Yakubiec died under a crush of coal.

Sir:-

I beg to advise you that on the 23rd inst., I investigated fatal accident to Frank Yakubiec, which occurred at 10.45 a.m. on the 22nd inst, while following his employment as a miner at mine # 80, operated by the Natural Resources Dept., of the Canadian Pacific Railway, Bankhead, Alta.

This fatal accident occurred in # 14 chute, 0000 seam, A level, north side, No. 80 mine.

According to Mike Perrotti, loader, who was the only person in the vicinity at the time accident occurred, he, (Mike Perrotti) was on the manway about 15 ft. from the counter gangway, when he heard coal running down the chute, and about the same instant heard someone shout from inside the chute down below him. Mike Perrotti descended to the Bulkhead and was given instructions from the deceased (F. Yakubiec) to let chute boards off, and let coal down on gangway. He ran off about two cars when deceased shouted for him to stop the coal from running. Frank Yakubiec, deceased, then ordered Mike Perrotti to get an axe and cut a hole in chute, which he commenced to do, when the motor man, Joe Travis, came along the gangway. Mike Perrotti told the motor man that a man was in the chute, and he had better go and get the fireboss and gangway men. The motorman went up to the bulkhead and heard deceased shout "Mike" several times, and then proceeded inside to get the fireboss and gangway miners.

Working in a coal mine in the early 20th century was risky — processes were dangerous and safety standards lax. Accidental deaths like Frank's were common.

Feast or Famine

I open my eyes to take in the precious daylight. "Grim" is the first word that comes to mind.

Standing at the lamp house and feeling its haunting ambience, I know there's no way I could have mustered the nerve to trade a piece of ID for a brass lamp and the possibility of a one-way trip into a clammy black hole.

I reflect on Frank's contribution to providing fuel that heated homes in the thick of winter or stoked steam engines for the pleasure of wealthy tourists riding a train through the Rocky

Mountains. Frank dug out a lot of coal so others could enjoy those things — and gave his life in that pursuit. Yet I'm pretty sure that anyone who enjoyed the benefits of coal from Mine No. 80 had no idea where their fuel came from, nor what sacrifices — including Frank's and many other miners' lives — were made under that mile of mountain.

I don't think men like Frank and Mike were looking for people to stop them on Main Street and thank them for risking their lives so others could have a warm bath. Just the opposite.

From what I've read, Rocky Mountain coal miners were especially stoic. Many were hardworking immigrants from Eastern Europe looking for a better life in Canada. I doubt they found an easier existence here, but I know they wanted to be paid fairly and treated well for the sacrifices they made — compensation and dignity make up for a lot of hardship.

Bankhead appears to have been one of the better mines in the area for working conditions, but, by today's Western standards, it was hardly acceptable. Tension between the United Mine Workers of America and the tweed-suited owners of the Pacific Coal Company led to many strikes over the mine's 20 years of operation.

Indeed, labor strife was a major factor leading to the town's demise. The miners wanted better pay and working conditions, while the owners stereotypically remained intransigent with their profits (or lack thereof).

Coal was a feast or famine business, increasingly the latter, because the cost of production in Alberta was too high relative to the volatile ups and downs of the commodity price. The downturns in the coal market were hard on company coffers and even harder on workers' wallets.

The final, irreconcilable strike began on April 1, 1922. For 76 days, workers refused to gather at the lamp house. Looking around these ruins, I can imagine union leaders and management reps negotiating for hours in a dingy room hazy with cigarette smoke. It must have been hard enough cutting through all that smoke and bullshit in the air, let alone cutting a pay deal.

So there was no deal.

On the evening of June 15, Bankhead began its abrupt decline from bustling, proud mining town to a historical curiosity for hikers.

Powered by compressed air, the dinky was much stronger than a horse and a lot safer than a locomotive making sparks in a gas-filled mine.

An Assault on All Fronts

A gust of wind rustles the leaves, breaking my train of thought and directing my attention to a path. Intrigued, I head over. I arrive at what the miners called a dinky, a hefty air-powered iron cylinder used to haul materials into and out of the mine. It looks like a prop from a steampunk show.

Not far from the dinky, leftover slags provide evidence of what's been written about Bankhead's mine: the quality of its coal was not that good. Sure, the stuff looks black, but it's full of rocks and too wet in composition, prone to crumbling into small grains. Fireplaces like to be fed dry, chunky coal, so, after the dinky hauled coal from the mine, workers had to pick out the rocks, which added a lot of manual labor.

As Mine No. 80 developed, its coal output became finer and finer, which meant the yield of desirable coal chunks was poor relative to the effort required. Bankhead suffered from the most serious of business maladies: poor-quality product and a high cost of production.

I start walking back down the path, my mind wandering farther than my feet.

The operation had challenges on many fronts: labor discontent, poor product and thin profits. Another competitive drag was that Bankhead was a thousand kilometers from the lucrative markets to the east, so transportation costs were prohibitively high.

Bankhead had big-picture business issues, but I also know that local circumstances played a part in taking the place down. The Pacific Coal Company was a sophisticated operation backed by the ample wallet of its parent, the Canadian Pacific Railway. Having corporate backbone was good, but in Alberta, there were plenty of "mom and pop" coal-prone sites. Collectively the little guys could produce fairly large quantities of coal without a lot of up-front money. I've been to some of those sites too, where a few picks, shovels, a donkey, some rickety carts and a cadre of small-time entrepreneurs were enough to get into the coal business.

Whenever the price of coal went up — as it usually did in the dead of winter — the little guys flooded the local market with too much of the stuff. Prices fell quickly in the spring and nobody, big or small, could make a living.

The historical record speaks to brutal competitive circumstances. Owners, managers and workers at Bankhead were constantly challenged to make a buck in this recurrently adverse micro-environment.

After World War I, change hurtled at the coal business like a freight train — literally. Diesel engines powered by oil succeeded coal-fired locomotives in North America. Lessened demand to fill a tender car with coal at pit stops like Bankhead further eroded the mine's sales. The contemporary parallel to this development is obvious: if everyone starts buying plug-in electric vehicles, thousands of gas station owners will be subject to a similar fate.

Hope Ain't No Strategy

I emerge from the woods at the final stop on my tour: what's left of the church on the hill.

There's not much here, just crumbling foundations wrapping around what used to be the basement, now overgrown with shrubs and trees. The central feature is a wide stone

After World War I, change hurtled at the coal business **like a freight train — literally.**

The affectionately dubbed "Stairway to Heaven" still stands strong. The church it led up to, however, is long gone.

staircase that's still in pretty good shape. Affectionately named the Stairway to Heaven, it rises about three meters above the foundations before ending with a sharp drop to the basement. There's no building awaiting, just a view of the heavens.

From the top of the stairs, I survey the ruins and imagine miners and their wives praying for renewed prosperity during those tough times. This place is a case study of life in so many dimensions.

During my MBA, case studies about bankruptcies were coldly analyzed in a sterile classroom. Yet an Excel spreadsheet and a stuffy lecture from a prof in a wrinkled shirt can't begin to capture the human element of business failure and community demise. This old church — where unemployed workers desperately prayed for more food on their table — gives a "distressed" corporate balance sheet entirely different meaning.

The first thing I noticed on my trip to Bankhead was a love story. Despite the hardship, I can sense a love of community nestled here in the beautiful Rocky Mountains. I leave saddened that this love affair lasted only 20 years.

Heading to my car, it occurs to me that my day job may not involve religion, but it is about analyzing business, which is a church of sorts: corporate leaders and their faithful flocks of investors revere making money. When that's not happening, there's much gnashing of teeth, wringing of hands.

I think about the presentations and lectures I've given to people in the energy business over the years. I've observed that, in tough times, especially when oil and gas prices are depressed, the first companies to suffer from competitive pressure are those not at the top of their game. In every case, that underperformance leads management and employees to a strategy of hope and prayer, rather than innovation and improvement. But business, unlike church, is an unforgiving environment, one where success, and failure, are indifferent to such folly.

I always see heads in the audience nod sagely at my common-sense conclusion: if you want to survive in the energy business for the long haul, you must have quality assets; a productive, innovative and disciplined organization that keeps costs low; and access to diverse and lucrative markets that aren't plagued by the ups and downs of seasons and other cyclical factors. I realize that isn't a message everyone wants to hear though, especially those in financial trouble.

Instead, maybe I should encourage my audiences to forget their spreadsheets, stop listening to me and plan a trip to Bankhead. Anyone who comes here will realize that merely hoping for business success can lead to their own Stairway to Hell — a place where too many have found their end.

WHEN STOKE STOKES

ENVIRONMENT

PREAMBLE

IN 1947, "GEORGE," MY protagonist, wrote a tender yet haunting message on a postcard bearing an industrial scene that borders on post-apocalyptic. He acknowledges the scene is bad, yet he *chose* that card to send to his daughter.

We'll never know what prompted George's selection, but I do know how such a grim image wound up on it in the first place. In post–World War II England, such an environmentally egregious sight signified industrial fortitude.

Of course, our values have since shifted and, with them, our visual indicators of prosperity. So, as you read this story, think about how our recalibrated values have affected such depictions. What *does* prosperity as a society look like now?

Climate change and other sustainability issues have necessarily pushed ecology to have a greater influence in the tensions between sustainability and prosperity. The tensions are palpable. Photos of wind turbines and solar panels elicit a welcome sense of progress, yet does such imagery reflect the reality of what's going on in the world? And back in the realm of fossil fuels, are grim images of pipes and smokestacks truly representative of every oil and gas operation?

While writing “When Stoke Stokes,” I was left wanting for answers to these and other sensitive questions. Parsing the best and worst of the world always leaves me seeking the middle ground. In the end, my box of postcards, a note from a father to his daughter and a god-awful photograph gave me a new perspective on what leaders in the energy business must do to express their vision of the future.

There's a perverse amplification here: **the dirtier the image, the louder the message about wealth and power.**

WHEN STOKE STOKES

GEORGE MISSED HIS daughter terribly.

During World War II, he and his wife, Clara, had sent Katherine overseas to Branksome Hall, a prestigious boarding school in Toronto, far from the Nazi bombers flying their destructive missions across the English Channel.

After the war, the family remained separated. Katherine continued her studies at Branksome, while her parents put their lives back together in England. Even though air raid sirens had gone quiet almost two years earlier, living conditions remained tough in England's industrial heartland.

Repairing British factories and restoring essential civic services were top priorities after Germany's surrender in 1945. Badly impaired by the bombings, the country's infrastructure to distribute fuels such as coal and petroleum was still hobbled. Labor problems exacerbated the challenges. So George and Clara knew that domestic comforts like reliable heating and

Energy usage and prosperity have always been linked. In 1947, as Stoke-on-Trent rebuilt its manufacturing infrastructure, these sooty emissions were emblems of a thriving economy.

lights would have to wait. Though, given the joy and relief they'd felt at the end of the war, the discomfort was negligible.

By early 1947, coal stocks in the United Kingdom were severely depleted. The hardship — dubbed the Coal Crisis — was aggravated by a winter so unusually cold and snowy it made international news. So, in a dimly lit home, George sat at his desk, pulled out a fountain pen and considered the postcard he'd bought from a street vendor.

His eyes were drawn to the ominous chimneys discharging smoke. Even the sight of it on paper seemed to make his eyes sting. It certainly wasn't pretty, but then, not much was in the aftermath of the biggest war the world had ever seen.

George's gaze drifted from the smoke in the picture to the fireplace opposite his desk. Lumps of coal hissed in its small cement cavity. He felt warmed by the sight, but shivered in demurral. Shaking his head, George returned to the postcard, neatly penning a note of reassurance to Katherine:

> It isn't quite as bad as this during the coal crisis but it's awfully cold.
>
> – LOTS OF LOVE, DAD

Bundled up in his Chesterfield coat and as much cable-knit wool as he could find, George ventured out into the damp morning on February 24, 1947. He reached the bright red pillar-box and, after checking that the two-pence stamp was affixed, posted his card. He spent the walk back imagining its long voyage across the Atlantic, a voyage he'd dearly love to take, if only to see his daughter again.

A Scene Only Dickens Could Love

Sixty years later and thousands of miles away, I found that postcard at an antique fair in western Canada, sandwiched in a shoebox among hundreds of other collectibles.

I like rummaging through old postcards like the one sent to Katherine, looking for imagery that reflects energy themes from yesteryear — old mills, ironworks, steam locomotives or hydrodams. If I'm lucky, people like the father I'm calling George have also penned sentiments to their acquaintances.

Let's be honest though: postcards are mostly for making family and friends jealous. It's the smug "Wish you were here!" scrawled over a sunny beach, misty Niagara Falls or the Eiffel Tower backlit by a Parisian sunset.

George's postcard is hardly a feel-good, envy-inducing image. When I first saw it, I figured he'd be saying "Wish I was there! Get me outta here!"

And what about the card's designer? Why would anyone *choose* to put this imagery on a postcard? Three chimneys in the foreground belch dark smoke from burning coal. In the background, dozens more spew soot. It's a grim industrial silhouette that suggests a dying city gasping for breath. What civic booster could think such a post-apocalyptic landscape would entice *anyone* to wish they were there?

Charles Dickens would have loved this photo. In his 1852 novel, *Bleak House*, the grand master of dreariness wrote of London: "Smoke lowering down from chimney-pots, making a soft black drizzle, with flakes of soot in it as big as full-grown snow-flakes — gone into mourning, one might imagine, for the death of the sun."

Judging by this card, the sun appeared to be long dead in 1947. Its dismal scene is not Dickensian London, but post-war Stoke-on-Trent, a centuries-old manufacturing center four hours northwest of London that helped birth the gears, levers and searing kilns of the Industrial Revolution. It's the type of place populated with endless rows of brick buildings, each full of gargantuan machinery and combustion chambers, any of which could have easily glazed a careless worker into blue-collar crockery.

The card's catchy slogan — "When Stoke Stokes" — makes me smile. I admire the cleverness of these three short words.

Why would anyone *choose* to put this imagery on a postcard?

Combined with the imagery, they conjure a scene of someone with a puffed-up chest, bellowing, "Gather 'round, gather 'round, ladies and gentlemen, boys and girls! Marvel at the impressive sight! Witness what happens when the pottery kilns, fireplaces, ironworks and coal-fired machinery of Stoke-on-Trent are all stoked up at once!"

This kind of industrial bravado was an anthropogenic theme that ran through the ferrous DNA of many European manufacturing towns like Stoke-on-Trent in 1947. But that still doesn't explain why anyone would put such an environmentally offensive image on a postcard.

Filthy Rich

Deltiology, or the study of old postcards, is, I admit, an eccentric pursuit. Yet it's great fun and offers a unique sense of what people were thinking about in bygone eras.

Putting the card aside, I flip through my energy-biased postcard collection. In each one, I note similarly unappealing

The more I look
at my collection,
**the more I see
a common
denominator
of economic
machismo.**

Symbols of the new age of transportation and industrial might, oil gushers were once glorified. This explains why this 1935 image was embellished with a plume gushing from the derrick.

industrial imagery — bizarrely commonplace from the late-19th to mid-20th centuries.

A gruesome 1910 postcard from South Africa gives me pause: whale carcasses transported en masse on rail cars for "refining" into lamp oil. Several from the same time period depict denuded forests, their Californian or Texan landscapes re-seeded with the dense wooden stubble of oil derricks. On others, smoke plumes look as if they've been manually smudged onto chimneys. In addition to emphasizing industrial operations, these sooty embellishments seem to accentuate, even glorify, the drama of air pollution.

The more I look at my collection, the more I see a common denominator of economic machismo. Energy systems based on fuels like coal, whale oil and crude oil gave muscle to the Industrial Revolution and beyond. For over 200 years, these fuels and factories created wealth. Making stuff on a grand scale represented economic power; therefore, gritty industrialization was a virtue.

In other words, the subliminal message is "Big things that make a mess reflect well on where we live — proof we have all the privileges of economic power and comforts of modern living." There's a perverse amplification here: the dirtier the image, the louder the message about wealth and power. Rather than it depicting a city's last gasp, this postcard was Stoke-on-Trent thumping its industrial chest.

From a 21st-century perspective, my thesis may be hard to swallow. But recall former US President Richard Nixon, who praised excessive energy consumption as late as 1973 in these remarks: "... there are only 7 percent of the people of the world living in the United States, and we use 30 percent of all the energy, in the United States. That isn't bad; that is good. That means that we are the richest, strongest people in the world and that we have the highest standard of living in the world. That is why we need so much energy, and may it always be that way."

This controversial president did sign major amendments to the Clean Air Act of 1970 and even formed the Environmental Protection Agency, but his energy sentiments were more in tune with the end of a long era where the pursuit of GDP, power and Western comfort trumped the environment.

I experienced energy consumption synonymous with smokestacks and tailpipes firsthand. As a kid in the 1970s, I grew up breathing leaded gasoline, sitting in the back of a big, buoyant Pontiac, exploring California's urban landscapes from its freeways. It was a great feeling of adventure, but my eyes burned from skies tinted with an orange-brown patina of smog. I was sensitive to the irritant, but because my parents didn't complain, neither did I. Thinking back on such complacency, I guess that they, like most others at that time, were duped into believing environmental degradation was a normal trade-off for the freedom of gas-guzzling mobility.

Perception Is Reality

As I pick up Katherine's card again, my thoughts about hidden energy messages still hang in the air.

I find it hard to believe George was making a veiled statement about industrial strength to his daughter. He probably just wanted to remind her of home and assure her that, despite the cold, all was well. Still, what an odd choice. His loving note contrasts sharply with that unloved (and downright hard-to-love) environment in 1947 England.

Flipping through all my industrial postcards, I remind myself that unsightly energy use was pragmatically glorified because it was seen as beneficial to the economy. Decades later, the way we portray and deliver the imagery is different, but one thing hasn't changed: the images we're presented with both reflect and influence our societal attitudes.

It occurs to me that social media is our modern-day postcard — dispatches from lands near and far flashing a personal perspective, if not an opinion. And when it comes to environmental debates about energy, social media is hugely influential. In fact, it's informing how our energy decisions are made and accepted — or, more often, rejected.

Whale hunting was a grim chapter in our history. Although whale oil began being replaced by petroleum in the 1870s, the hunts continued into the 1900s.

Yet how true is the message depicted by a photograph? Was George's postcard an authentic representation of Stoke-on-Trent? He answers that himself: "It isn't quite as bad as this . . ." So, is one photograph of a dirty oilfield indicative of the business at large? Or is a wind farm in a bucolic field really as idyllic as it appears?

Energy leaders seeking to power our future, whether by fossil fuels, renewables or whatever's on the horizon, must always remember they're in the image business. People *will* judge a book by its cover, and how an energy company visually presents its operations translates into perceptions — true or not — that shape public opinion and even government policy.

Putting my postcards away, I smile, realizing their message runs much deeper than "Wish you were here."

LONG WAY AROUND

PREAMBLE

I GREW UP IN THE 1960S, when power outages during winter were common. When the lights went out, my mother would matter-of-factly bring out the candles. Nobody in my family panicked. That's because we had some sense of energy security, simple as it was, until the power came back on.

In "Long Way Around," I explore energy security at a national level. It's a story about sudden oil shortages in Western countries due to the 1973 Arab Oil Embargo. I remember that era of Middle Eastern geopolitics — unlike my mother, many governments were unprepared for the energy shortages that ensued.

In 1973, Canada was rich in oil and gas but lacking in self-sufficiency. So the country had to hire tankers to make a circuitous journey, from the Pacific Ocean to the Atlantic, carrying emergency supplies from one side of the country to the other.

My story about one of those tankers, the Kimon, is obscure. Yet the imagined tale of its captain and crew gives insights into the trade-offs societies make when building out their energy infrastructure. In balancing cost, environment, economy and security, the last is often overlooked during times of stability.

Canada is still rich in oil and gas… and it still isn't self-sufficient. To this day, oilfields in the west are not connected to the large consumer base in the east. "Long Way Around" reminds us that we tend not to think about energy security until the lights go out.

Pipeline debates in Canada are nothing new, nor are the reasons behind them.

LONG WAY AROUND

THIS IS A story about energy security. It's about what happens when the amenities a nation depends on are abruptly threatened — or cut off altogether.

If you lived through the 1970s like I did, you probably know what I'm referring to: sudden gasoline shortages and panic at the pumps framed against the never-ending geopolitics of the Middle East. In short, a fast-breaking situation when our taken-for-granted energy supplies became scarce and uncertain.

Yes, 45 years after the Arab Oil Embargo, I'm reminiscing about energy shortages.

My musings are prompted while standing on the north shore of the Burrard Inlet in beautiful Vancouver, Canada. Across the inlet are the loading docks at the end of the Trans Mountain Pipeline, where the oil tanker Aristoklis awaits.

It's summer 2018. I've come here because it's the day the Canadian government nationalized the 65-year-old pipeline. And although prickly environmental issues primarily drove the decision to bring the pipeline into public ownership, a more interesting story echoes through this steel tube — about how society prioritizes security in the energy choices it makes.

Same Country, Other Side

As its name implies, the Trans Mountain Pipeline brings oil to this tidewater by crossing two Canadian mountain ranges: the Rockies and the Cascades. It's a thousand-kilometer westward journey from Alberta's prolific oilfields to British Columbia's coast, where I'm standing.

As I ponder the massive vessel across from me, my imagination wanders back to November 1973 and what must have been a similar scene...

The Kimon, a Greek tanker, was moored at the same spot I'm looking at now. The well-traveled ship's captain — I'll call him Nick — had docked uneventfully. Special loading arms that looked like giant, kinked drinking straws were filling the Kimon from storage tanks on shore. The reservoirs were presumably well stocked, having been topped up with crude oil flowing in from the pipeline.

Meanwhile, Captain Nick was on the bridge planning his ship's 14,000-kilometer route — a distance greater than a third of the way around Earth. Fidgeting with his curly white mustache, he looked up at Alex, his first mate, and delivered his assessment. "We'll be at sea for three weeks."

"It's a long way, sir," Alex observed, "but we're close to land for most of it... should be a calm journey."

Looking over the maps, Alex followed Nick's carefully ruled navigation lines. They would route through the Strait of Juan de

The 30,000-ton MS Kimon was the first of several tankers to transport Canadian oil during the Arab Oil Embargo. In one month alone, they moved 1.5 million barrels. *Photo courtesy of City of Vancouver Archives; CVA 447-5455, photographer Walter E. Frost*

Fuca, out to the eastern edges of the Pacific Ocean, sailing south to the Panama Canal, then back up through the Gulf of Mexico and along the Atlantic coast.

"Helluva long way," Nick muttered, "for what could have been a direct overland haul."

Alex pulled a devilish grin. "At least it keeps us working, Captain."

"Yeah," Nick agreed as he leaned over his maps, tracing a finger along their final destination: the Saint Lawrence Seaway.

"That's where we end up" — the captain tapped his finger firmly on the spot — "Montreal. Right back in Canada, only, on the other side."

In the Grip of a Crisis

As a Greek national, Nick may not have fully appreciated the importance of his journey, but I can tell you that the Kimon's voyage was urgent. Eastern Canada needed fuel from western Canada's oilfields as soon as possible.

A month prior, Arab countries had united to express their anger at the Western world. Unhappy with supporters of Israel in the Yom Kippur War, oil-rich countries like Saudi Arabia, Syria, Egypt and Libya turned the valves down on Canada, the United States, several European countries and Japan. The restrictions were tight enough to cause energy havoc, and anxiety was radiating around the world. Oil flows were being choked off and Western fuel supplies were running short.

Hostage to the geopolitics, and their dependency on their angry oil suppliers, the sanctioned countries had to fend for themselves. The price of oil soared. National hoarding of petroleum products took hold. Each country had to secure its own alternative sources of the fuel that was needed for turning wheels on every axle of the economy.

International relations were cold. So was the weather. In the Northern Hemisphere — where most of the target countries were located — winter was setting in and fuel for heating homes, generating electricity and gassing up the family Buick was getting tight. The shortages of oil were forcing countries to take extreme action.

That's why the Kimon's journey was so absurd, yet so urgent. Although the distance between Vancouver and Montreal is only about 3,600 kilometers as the crow flies, Canada lacked the means to transport the oil directly via land. The sole west-to-east oil pipeline at the time was the Interprovincial Pipe Line

THE VOYAGE OF THE MS KIMON

And Its Cargo of Canadian Oil

In the absence of a direct pipeline, this 14,000-kilometer, three-week-long journey helped mitigate the effects of the oil crisis in eastern Canada.

system, but it went only about halfway, to Sarnia, Ontario. And its capacity was limited.

So the Kimon was set to embark on a trip four times longer than it needed to be — to deliver a couple hundred thousand barrels of oil back to the country it came from.

My mind returned to the bridge of the Kimon.

Precious Cargo

"This oil's worth a lot of money now," mused Alex. "I heard the price jumped to over four dollars a barrel yesterday. That means..." He did some quick math and made an impressed whistle. "It's worth almost a million dollars."

The Kimon's crew appreciated the value of their ship's cargo, but as international sailors on a Greek-registered vessel, they were indifferent to Canadian energy issues. Their job was to get the cargo from point A to point B.

Nick nodded, but his mind was elsewhere. "How is it," he said, scratching the back of his neck, "that a country so rich in oil is not self-sufficient in this stuff from coast to coast?"

"Don't know." Alex shrugged, peering through binoculars at the deck hands working below.

Nick shook his head. "I'm sure our fellow countryman Aristotle would have had something to say about the matter. When he coined *enérgeia*, he used it in the sense of being at work." He added wryly, "Canada's great energy supplies are hardly working for it *now*."

Philosophy aside, Nick was just happy that Canada's unfortunate situation was contributing to his livelihood.

In truth, Canada once tried to become self-sufficient in oil from west to east. But the politics were contentious, so a Royal Commission on Energy was dispatched to study the matter. In 1959, its report rejected building an Edmonton-to-Montreal pipeline. The reason? In large part because overseas oil — much of it from sanctioning producers like Saudi Arabia, Libya and Iraq — was deemed plentiful. Buying from foreign sources was cheaper than laying a metal pipe across a few thousand kilometers of prairie, forests and granitic rocks. Basically, cheap oil from other countries trumped the hassle and incremental cost of securing domestic oil.

An hour passed on the Kimon's bridge. Just as Nick was finishing his paperwork, Costas, the communications officer, delivered a note. "Telegram, sir." He knew the news wasn't good.

Reading the message, Nick twisted his mustache in irritation. "This is crazy! I was assured these government types were going to set aside their trade protection rules in time of national crisis!"

The message was simple:

DELIVERY TO MONTREAL NOT POSSIBLE. STOP.
RE-ROUTE TO PORTLAND, MAINE, USA. STOP.

The telegram went on to advise the Kimon that it would not be able to unload its cargo in any other Canadian port. Cabotage rules — prohibiting foreign-owned ships from hauling cargo between two domestic ports — would not be waived, not even in the midst of geopolitical crisis.

Instead of a direct sea voyage, Alberta's oil would have to be unloaded in the nearest American port and flow to Montreal via an American pipeline.

Cheap oil from other countries trumped the hassle **and incremental cost of securing domestic oil.**

"Alex," Nick said firmly, "give me the east coast maps of the United States. Change of plans. We're heading to Maine."

At What Cost?

Records show the Kimon and its 200,000 barrels of oil arrived in Maine just before Christmas 1973. But there was no time to celebrate. Due to the escalating sanctions, the tanker had to return to Vancouver as soon as it finished unloading. More Alberta oil awaited.

In December alone, the Kimon and seven other tankers — all foreign-owned — transported 1.5 million barrels of oil from Vancouver, through the Panama Canal, to the northeastern United States, where a pipeline awaited. Their circuitous journeys helped mitigate Canada's energy crisis, though with anxiety and at tremendous cost.

I take photos of the dock, turn around and walk back to my car, as perplexed as Captain Nick. Forty-five years after his voyage, Canada would still be in the same insecure situation should another energy crisis hit.

Too Little, Till It's Too Late

The events leading up to the 2018 Act of Parliament that nationalized the Trans Mountain Pipeline were contentious, but the reasons can be distilled down to intense public debates about environment and safety versus economy and prosperity. Energy security was a forgotten memory.

Pipeline debates in Canada are nothing new, nor are the reasons behind them. Political polarizers from past decades include the TransCanada Mainline (1954), the previously mentioned Edmonton–Montreal Pipeline (1957), the Mackenzie Valley Pipeline (1974), Northern Gateway (2005), Energy East (2013), Keystone XL (2016) and the proposed expansion of the same Trans Mountain Pipeline (2018) from which Nick loaded his ship.

In many ways, Canada is a laboratory that has conducted economic experiments on trade-offs in its energy supplies. Yet what I've gleaned from my reviews is that none of those experiments has articulated energy security as a primary objective. Consequently, this country's energy needs are no less vulnerable to geopolitical forces than they were decades ago. That's not an indictment, it's an observation about priorities.

There's a recurrent theme in the stories I tell: people want their energy to be cheap, clean, safe and secure. In developed countries, that want has become an entitlement. Of course, we'd all like free, completely safe and secure energy that has zero environmental impact! Yet we don't live in a utopia, so compromises must be made based on social circumstance. Reflecting on Canada's pipeline debates, I sense that, since World War II, the urgency of energy security in the Western world has gradually lost rank relative to the cheap/clean/safe trifecta.

Captain Nick's story is important because energy security is a difficult dimension to value in the present. As long as electricity is always ready at the flick of a switch and the service station always has a nozzle to fill their car, it's not something consumers think about. Skimping on energy security is like neglecting house insurance — we don't think much about it until somebody robs our home . . . then it's too late.

Energy security is a difficult dimension to value in the present.

As I get in my car, I take for granted that the engine will start and I'll get to my destination without having to worry about where my next tank will come from. So too will the billion-plus people who drive oil-dependent cars in this world.

But this isn't just a story about oil. Electric vehicle drivers aren't concerned about where the charge in their batteries comes from either. Nor will you worry whether the light will turn on the next time you flick the switch. That's the way it should be — if we plan our energy systems appropriately.

Epilogue

Back at home, I research the Kimon in greater detail.

In September 1978, the Kimon had again been hired to mitigate an energy security deficiency. This time, the tanker found herself in trouble in the Mediterranean Sea, off the coast of Lebanon, where war had broken out. Beirut was cut off from oil supplies and needed the vital commodity, similar to Montreal five years earlier. Only, this mission was dangerous: it had to skirt warships and smuggle oil into the needy country. Not surprisingly, the tanker was seized and her crew arrested.

Released soon after, the crew and its ship enjoyed only short-lived freedom. Ensnared in continuing Middle Eastern conflict, the Kimon was sunk by rockets before the calendar turned to 1979.

In times of peace, we tend to discount the value of energy and take our sources of supply for granted. But during conflict, energy is often the first point of political leverage, and in extreme cases of military intervention, we pay the price with scarcity, hardship and potentially even loss of life.

Societies that learn to prioritize energy security when it's not an issue will never have to think about it when it becomes an issue.

ONCE UPON A TIME...

SOCIAL AND ETHICAL

PREAMBLE

As my wife and I raised our sons, we didn't realize we were inadvertently teaching them energy and environmental economics — just by reading them children's books.

Only years later did I look back and fully appreciate that classic stories from Virginia Lee Burton, Dr. Seuss and Bill Peet were making pointed statements about pollution, reckless consumption and energy transitions.

With my interest piqued by energy-themed children's stories, I bought a rare 1924 collected edition of three Hilaire Belloc classics, *The Bad Child's Book of Beasts, Together with More Beasts for Worse Children and Cautionary Tales.* Inside, I was quickly drawn to "The Whale." Although the poem is a pithy 71 words, Belloc's wit takes jabs at a wide target of energy and environmental issues, all against a background of colonialism, income stratification and the ethics of slaughtering animals to fill lanterns with lighting fuel. It hardly sounds like a story for six-year-olds!

In "Once Upon a Time...," I reflect on children's stories like Belloc's, along with Burton's *Mike Mulligan and His Steam*

Shovel, Seuss's *The Lorax* and Peet's *The Caboose Who Got Loose*. What influence did they have on children who grew up during a period of heightened environmental pressures?

This story reminds me of the truism that our future leaders are influenced at a young age. Then they grow up and become decision makers. That's why it takes a generation or two before societies realize meaningful change.

Put another way, a review of bedtime stories about make-believe worlds made me realize that generational attitudes toward energy and environment have to transition first. Only then, in our real world, can we expect tangible, physical transitions.

Never discount the sophistication of children's books **in delivering a message.**

ONCE UPON A TIME...

IF YOU'VE BEEN part of a child's bedtime routine, you know the joy of cuddling with them to read a favorite story. But did you ever think that this quiet ritual could shape their future beliefs about energy?

In my family, we accumulated lots of children's books over the years: fact, fiction and fantastical. For my wife and me, there was no greater satisfaction than seeing our boys soak up a good book.

Now, I'm no child psychologist, but from my experience, it's clear that a child forms ideas about what's "normal" based on what's read to them. Aristotle backs me up on this. He was the first to suggest that a young mind is like a blank slate that gets filled in with ideas and experiences over time.

I remember our sons' unblinking eyes poring over every illustration, and their unfiltered ears clamping on to every word, trying to make sense of a story's world. Gradually, the pages gelled into normative beliefs that would inform their lives (Santa Claus and the Easter Bunny excepted).

Long after they left the nest, I reflected again on the influence of children's books on adult beliefs. Take *The Jungle Book*, for example. At face value, Rudyard Kipling's classic shows kids that tigers, panthers and snakes are dangerous. Bears, on the other hand, are warm, fuzzy and friendly. After seeing Mowgli snuggling with Baloo, what child would think a bear is threatening? Other anthropomorphically inclined bear stories — like *The Berenstain Bears*, *A Bear Called Paddington* and, of course, *Winnie-the-Pooh* — instill a belief that bears are more likely to hug you than take a swipe at you.

Based on bear stories alone, you might think children's books are responsible for nurturing grown-up idiots. You know the people I'm talking about: the ones who stop their cars on highways to feed half-eaten peanut butter sandwiches to grizzly bears. Maybe they do it because their parents read them *The Jungle Book* too many times. That's just my theory, but think about it: do you know *anyone* who stops their car to offer Doritos to a snake? (I obviously like to keep my philosophy lighthearted.)

In Search of a Happily Ever After

I have a rule to not talk or think about my work in two sacred places: the bedroom and the dinner table. So energy wasn't a subject I pushed on my family.

Now, years later, I'm trying to recall the books that seeded a certain energy mindset, subliminally or otherwise, into our children. I can think of a few.

"So you see, Alex," I might have said after reading *Thomas the Tank Engine* to my eldest, "locomotives like Thomas don't use steam and coal anymore. Isn't that an interesting energy transition, son?"

Come to think of it, Alex had probably already learned about that transition from Virginia Lee Burton's classic *Mike Mulligan and His Steam Shovel.* She published the book in 1939, when soot-spewing, steam-powered locomotives, tractors and backhoes were still common. Real-life encounters with the clanging machines were undoubtedly softened by the image of Mary Anne, the personified steam shovel operated by her easygoing owner, Mike.

Reading Burton's tale again reminds me of Alex's angst when my hand turned from page 13 to 14 and I read the ominous words: "Then along came the new gasoline shovels and the new electric shovels and the new Diesel motor shovels and took all the jobs away from the steam shovels."

Energy transitions, corporate greed, environmental desecration — these classics gently introduce children to some very adult subjects, laying the foundation for their future attitudes toward energy.

Against this dystopian backdrop, the emergence of social conscience in children's materials is hardly surprising.

A look of sympathy overtook his face when he saw Mary Anne being spurned, much like the nerd on the playground once the cool kids show up. While the newfangled shovels get all the jobs, Mary Anne's kind are sold for junk or abandoned altogether. "No one wanted Mike Mulligan and Mary Anne any more."

Unbeknownst to Alex, he was learning an economics lesson on how technological obsolescence displaces labor and capital. His teacher? An endearing steam shovel. Never discount the sophistication of children's books in delivering a message.

Environment is the theme in two 1971 classics: Theodor Seuss Geisel's *The Lorax* and Bill Peet's *The Caboose Who Got Loose*. I certainly remember the '70s — a time of peak pollution and smog in Western cities. The degradation of our environment was captured in headlines bemoaning things like acid rain, ozone depletion and mercury ingestion. Incidents like those at Minamata Bay, Love Canal and Three Mile Island, along with predictions of the death of freshwater life in the Great Lakes, heightened awareness.

Against this dystopian backdrop, the emergence of social conscience in children's materials is hardly surprising. Influencing children was necessary for creating a generation of leaders who would address these disturbing problems.

Dr. Seuss's message in *The Lorax* isn't simply that reckless energy use is the root cause of environmental degradation. At a higher level, he evangelizes the connection between the corporate elite's greed and our planet's desecration.

Seuss's pointed narrative hasn't lost its edge over the years: energy production, especially the fossil fuel type, is invariably tangled up in missives about money-lust and destruction of the environment. But his original message directed responsibility to his impressionable audience. "UNLESS someone like you cares a whole awful lot, nothing is going to get better. It's not," the Once-ler — the one whose greed has wreaked havoc on the environment — tells a young boy.

Personally, I'm partial to Bill Peet's story about sweet Katy Caboose. Her lot in life is bringing up the rear, "with an endless black cloud of smoke rolling past." To heighten the effect, Peet illustrates Katy being pulled by a coal-fired steam engine, although, by 1971, most locomotives were diesel-powered. Katy is getting gassed nonstop, even when docked in the freight yard, where soot from the plant looming in the background joins smoke pumping from the trains.

Katy's fortune changes when her coupling breaks loose on a steep mountain bend. My kids instinctively took a deep breath when they saw Katy escape her environmentally insensitive master. As a runaway, she jumps the tracks, flies through the air and lands comfortably wedged between two evergreen trees. Birds and squirrels greet Katy, and she finds contentment in the serenity of the clean mountain air.

On the final page, Peet soothes his young audience with these words: "And how Katy did love it up there in the trees." As the book is closed for the night, both parent and child can go to sleep assured the world has a chance of becoming a better place.

The Tale of the Poor Whale

After I'd read the last bedtime story to Andy, my youngest, I didn't buy another children's book for many years. Until one day, as a collector of antiquarian books, I was alerted to a 1924 collected edition of three Hilaire Belloc classics: *The Bad Child's Book of Beasts, Together with More Beasts for Worse Children and Cautionary Tales.*

These cheeky pieces of satire are as much for adults as they are for children, although Belloc makes it clear who his target is. In the introduction to *Bad Child's Book of Beasts*, originally published in 1896, he states, "The Moral of this priceless work / (If rightly understood) / Will make you — from a little Turk — / Unnaturally good."

He sets out to do this by prompting children to think about irresponsible human interaction with nature. On the surface, he, along with illustrator Lord Basil Temple Blackwood, offers a whimsical portrayal of subservient animals in an arrogant, colonial, human-dominated world. But it's hardly Winnie-the-Pooh. Aristocrats hunt hippopotamuses with "bullets made of platinum" and slay a giraffe "with stones and sticks and guns and slings."

What caught my attention, and why I bought this rare book, are the five pages titled "The Whale," characterizing the grim standing of the marine mammal in our 19th-century energy supply.

The Whale that wanders round the Pole
 Is not a table fish.
You cannot bake or boil him whole
 Nor serve him in a dish;
But you may cut his blubber up
 And melt it down for oil,
And so replace the colza bean
 (A product of the soil).
These facts should all be noted down
 And ruminated on,
By every boy in Oxford town
 Who wants to be a Don.

It's hardly Winnie-the-Pooh.

I've written a lot about the whaling industry, which was big business — and bad news for whales — until the latter half of the 19th century. By the end of the 1800s, as many as 300,000 sperm whales, the principal prey at the time, had lost their lives so people could light their homes.

Thankfully, whales were spared when petroleum was discovered in commercial quantity in the 1860s and "rock oil" quickly replaced whale oil in lanterns. Until that moment, cutting up a sperm whale yielded a prized fuel that lit big cities like London and New York.

Belloc informs the wide-eyed child that whale oil can commercially displace the oil from a colza bean (or canola), a millennia-old sustainable biofuel also used for lighting. At face value, the suggestion is clear: harvesting whales for oil is a better business proposition than harvesting bean crops.

But Belloc is being cynical. His real message is that destroying this great mammal is not only unsustainable but reprehensible, given the availability of sustainable fuel alternatives like vegetable oils.

There it is again: that subliminal lesson on energy economics and how our fuel choices, old or new, can have negative consequences. Fortunately for Mary Anne, those consequences are short-lived — her steam power is put to another use and she lives

The perverse message of this illustration accompanying "The Whale" is that these majestic beasts had only one purpose: to serve humans' selfish need for comfort.

happily in her new life. Belloc's whale, however, is a victim of our insatiable appetite for comfort and convenience.

I was struck by the mentoring at the end of "The Whale." Children who aspire to positions of influence should "ruminate." Good advice from Belloc, I thought: our future leaders *should* be encouraged early on to develop critical thinking about energy, environment and economy — and to consider the consequences of their choices.

Still, it takes time to effect change. After Belloc published *The Bad Child's Book of Beasts*, several generations of children's books and many graduating classes of young leaders went by before the 1986 moratorium on commercial whaling. Even then, the National Marine Fisheries Service has estimated that almost 3 million whales were killed between 1900 and 1999. It's enough to make anybody, never mind a young child, cry.

By the 1900s, technology had given humans a distinct advantage over whales: this steamer fired a harpoon equipped with hooks and an explosive grenade that incapacitated the whale.

From Small Moments Come Big Change

As I contemplate the children's books spread across my desk, an important truth about the future of our energy-environment circumstances comes to me: societal attitudes must transition before we can expect physical transition.

I run a hand over the cover of my well-worn edition of Belloc's classics, take in its old-book smell and think about the parents who bought it almost a century ago. Snuggled in a comfy bed and reading it to their children, they were influencing the youngsters' worldviews, whether they intended to or not.

I wonder what I'll be reading to my grandchildren. When that day comes, I'll stick to my rule about leaving business (and probably bears now too) out of the bedroom, but I'll be much more aware of the big role those seemingly small moments can play in influencing their beliefs. I hope those values will prompt their generation to make smart, sustainable choices... and that they'll live a happily-ever-after energy future.

SOURCES CITED

INTRODUCTION

Page 4: "The International Energy Agency estimates..."
International Energy Agency, "Sustainable Development Goal 7: Ensure Access to Affordable, Reliable, Sustainable and Modern Energy for All"
bit.ly/iea_electricity

STAIRWAY TO HELL

Page 113: "More than 85% of the world's insatiable needs..."
BP, BP Statistical Review of World Energy 2017
bit.ly/bp_energy

ONCE UPON A TIME...

Page 168: "By the end of the 1800s..."
National Marine Fisheries Service, "Emptying the Oceans: A Summary of Industrial Whaling Catches in the 20th Century," 2014
bit.ly/mfr_whaling

Page 169: "Even then, the National Marine Fisheries Service has estimated..."
National Marine Fisheries Service, "Emptying the Oceans"
bit.ly/mfr_whaling

- CREDITS -

All objects shown are in the collection of Peter Tertzakian, with the following exceptions:

Original Discovery Oil Company, Limited Prospectus (page 56) from the collection of Glenbow Museum, Calgary, AB, Canada.

Report to the Old Court House, Calgary, AB, Canada (page 118, foreground) from the archives of Parks Canada.

Mike Perrotti's witness statement (page 118, background) from the Provincial Archives of Alberta.

MS Kimon (page 147) courtesy of City of Vancouver Archives; CVA 447-5455, photographer Walter E. Frost.

Photography and photo-composition by Peter Tertzakian: pages 25, 56, 64, 67, 78, 88, 101, 105, 113, 118, 120, 124, 163. Other images scanned from sources noted in the stories.

ACKNOWLEDGMENTS

AFTER WRITING MY last book, I promised I would never write another one because of the patience it saps from my family. But I did it again, in even more time-consuming fashion. I am ever grateful to my wife, Jan, and our two boys, Alex and Andy, for their understanding and encouragement in allowing me to follow my passion.

Many friends and colleagues read my stories and provided valuable feedback. Thanks to all of them for their time and candor.

Finally, my stories are not possible without the dedicated help of Lori Burwash. As managing editor of this book, and the Energyphile project as a whole, she works tirelessly to ensure a level of quality that exceeds even my lofty expectations.

ABOUT THE AUTHOR

THE QUINTESSENTIAL ENERGYPHILE, Peter Tertzakian has devoted his career to energy, first as a physicist, then as an economist and investment executive. He's written two bestsellers — *A Thousand Barrels a Second* and *The End of Energy Obesity* — and is sought around the world as a trusted, engaging speaker. Energyphile is the culmination of his passion and knowledge.

CPSIA information can be obtained
at www.ICGtesting.com
Printed in the USA
LVHW090752301119
638774LV00004BA/4/P